*******MILESTONES CO

TID BECOMES ENGINEERING TECHNOLOGY DIVISI

"ENGINEERING TECHNOLOGY EDUCATION STUDY" BY GRINTER/DEFORE PRESENTED ❏

ENGINEERING EDUCATION PUBLISHES FIRST "TECHNOLOGY EDUCATION COMMENTS"

COLUMN **1975**   FEATURES

ENGINEERING   F.V.   ISTRY EDUCATION

CONFERENCE   ISTITUTE MEETING

HELD **1977**   *ICAL EDUCATION*

NEWS ❏ "TEC   *ERING EDUCATION*

TEMPORARILY   ATIVES **1980** ETD

ESTABLISHES   ORK COMMITTEE ❏

"TECHNOLOG   *EDUCATION* **1983**

LAST "TECHN   IN *ENGINEERING*

EDUCATION 1   ISSUE PUBLISHED

1985 ETD   S ENGINEERING

TECHNOLOG   NORTH INSTITUTE

OF TECHNO   *NG EDUCATION*

PUBLISHED   IG TECHNOLOGY

EDUCATION   ENGINEERING

TECHNOLOG   GY TEN-VOLUME

COMPENDIU   *TECHNOLOGY*

INSTITUTION   BERGER AWARD

PRESENTED (RS OF SERVICE TO

THE ENGINEERING AND ENGINEERING TECHNOLOGY COMMUNITIES ❏

Engineering Technology
AN ASEE HISTORY

Engineering Technology
AN ASEE HISTORY

General Editor
Michael T. O'Hair
Purdue University Programs, Kokomo

Manuscript Editor
Marilyn A. Dyrud
Oregon Institute of Technology

Production Editor
Barbara A. Wolf
Klamath Falls, Oregon

Managing Editor
Lawrence J. Wolf
Oregon Institute of Technology

Engineering Technology Centennial Committee
Alexander W. Avtgis, Wentworth Institute of Technology
Jack Beasley, Purdue University Programs, Anderson
Stephen R. Cheshier, Southern College of Technology
Frank A. Gourley, Jr., West Virginia Institute of Technology
Ann Montgomery Smith, Wentworth Institute of Technology
Anthony L. Tilmans, Southern College of Technology
G. William Troxler, Capitol College
Robert J. Wear, Academy of Aeronautics (retired)

McGraw-Hill

LIBRARY
ST. LOUIS COMMUNITY COLLEGE
AT FLORISSANT VALLEY

This book is dedicated to the technology students and instructors of the twenty-first century.

Cover/Interior Design: Frank Stanton
                             St. Louis, MO

Glencoe/McGraw-Hill

Editorial Director: Richard P. Reskow
Executive Editor: John J. Beck
Production Manager: Cindy Brunk
Production Editor: Jan Hall
Manufacturing Manager: Mark Bourgea
Design Director: Bart Hawkinberry
Cover Design: Brent Good
Copyeditor: Deborah Payne

Library of Congress Cataloging-in-Publication Data

```
Engineering technology : an ASEE history / general editor, Michael T.
   O'Hair; manuscript editor, Marilyn A. Dyrud; production editor,
   Barbara A. Wolf; managing editor, Lawrence J. Wolf; Engineering
   Technology Centennial Committee.
      p.   cm.
   Includes index.
   ISBN 0-02-800403-5
   1. Engineering --Study and teaching--United States--History.
2. American Society for Engineering Education--History.  I. O'Hair,
Michael Thomas, 1944-   .  II. American Society for Engineering
Education.  Engineering Technology Centennial Committee.
T73.E475   1995
620'.0071'073--dc20                              95-12447
                                                 CIP
```

Engineering Technology: An ASEE History

Copyright © 1995 by Oregon Institute of Technology, Klamath Falls, Oregon. Published in joint cooperation with Glencoe/McGraw-Hill. All rights reserved. Except as permitted under the United States Copyright Act, no part of this publication may be reproduced or distributed in any form or by any means, or stored in a database or retrieval system, without prior written permission of the Oregon Institute of Technology, Klamath Falls, Oregon.

Glencoe/McGraw-Hill
936 Eastwind Drive
Westerville, OH 43081

ISBN 0-02-800403-5

Printed in the United States of America

1 2 3 4 5 6 7 8 9 RRD-C/MC 02 00 99 98 97 96 95 94

Contents

Preface ix

Acknowledgments xi

List of Abbreviations xiii

Introduction 1

Chapter 1: The Presence of the Past 7

    In a Few Minutes 7
    Division Development and Council Conception 9
    Projects in Perspective 11
    At Issue Here 19
    Committed Committees 22
    Lead On 24
    A Little R and R (Researching and Reporting) 24
    ETD Newsletter 31
    A Little A and A (Affiliating and Articulating) 31
    Service with a Smile 32
    References/Additional Resources 32

Chapter 2: History of the Engineering Technology Leadership Institute 37

    Maiden Voyage 38
    More Planning 39
    NFET and ETLI 41
    New Directions 42

Chapter 3: Journal of Engineering Technology 45

The Early Development 45
The Inaugural Issue 49
Ten Years of Publication 52
Conclusion 56

Chapter 4: The James H. McGraw Award 59

Chapter 5: ETD/ETC Minutes 103

Part I: Excerpts from Division Meetings,
  1949-1992 104
Part II: Excerpts from Council Meetings,
  1964-1993 153

Chapter 6: Oral Histories 193

Walter M. Hartung 196
Lawrence V. Johnson 202
Michael C. Mazzola 212
Hugh E. McCallick 220
George W. McNelly 236
Winston D. Purvine 258
Richard J. Ungrodt 270
Eric A. Walker 292

Appendices 303

Appendix A: ETD Name Changes 304
Appendix B: *TEN* Reprint 305

Appendix C: ETD/ETC Officers 307
Appendix D: ETD Newsletter Data 317
Appendix E: ETLI 329
Appendix F: *JET* Data 330
Appendix G: ETD Meeting Minutes 332
Appendix H: ETD Membership Data 336
Appendix I: Captions and Credits, Photos 340

Index 353

Preface

A book? That's certainly not what the Engineering Technology Centennial Committee initially considered as the vehicle for a history of engineering technology education to celebrate the American Society for Engineering Education's centennial year. A short history was but one aspect of the committee's centennial plans, which included sponsorship of two regular sessions, a poster session, and a mini-plenary session related to the history of ET education, all slated for presentation at the 1993 ASEE Centennial Conference, Champaign-Urbana, Illinois.

Two years ago, when the committee embarked on this project, it was first conceived as several short articles to be published in some future issue of the *Journal of Engineering Technology*. As deadlines and time passed, it became a stand-alone monograph. Then a longer monograph. And finally, when the complexity of the project became clear as pages and pages of real text emerged in January 1993, the committee started referring, in solemn, hushed tones, to THE BOOK.

To narrow the focus, committee members decided to limit the scope to engineering technology education within ASEE, rather than chronicle a history of the entire movement — something, incidentally, that has been mentioned in ET business meetings since 1965. References are made to the Accreditation Board for Engineering and Technology, Tau Alpha Pi, and the history of individual educational institutions, but the focus remains on ET history within ASEE.

Although every attempt has been made to be as accurate as possible, committee members recognize their limitations. We are engineering technology educators, not professional historians. This book thus includes, in addition to well-documented events, oral histories from pioneers in the field, anecdotal accounts, and some "best guess" recollections. There are undoubtedly errors of omission of both events and individuals, and for that we apologize. We also made an overt decision to make the history interesting to read, perhaps to the detriment of scholarship. Even with these limitations, we believe this work will be useful to current and future scholars who study this discipline we call engineering technology education.

The oral histories contained in Chapter 6 represent not only an important part of this book but were a significant resource for the authors who prepared the introduction and initial five chapters. The oral histories may prove to be the most valuable and interesting portion of this project. Recollections of these ET pioneers help us to fill in the gaps from the few available source documents.

All of the volunteers involved in this project deserve a special thanks from the engineering technology community. I particularly want to thank the members of the Engineering Technology Centennial Committee for their two-year commitment.

As with any project of this scope, some individuals distinguish themselves by their service. Frank Gourley took on the challenge of writing the history of the Engineering Technology Division and the Engineering Technology Council. His attention to detail reflected in these chapters deserves a special thanks. And for those of you who have volunteered as manuscript editors, you know how much work is involved in preparing a manuscript such as this one. Marilyn Dyrud worked many hours editing it and putting it on a desktop publisher. She has also been an inspiration to me and other committee members. As the only English professor on the committee, she made us technologists look good. Thank you, Marilyn.

Michael T. O'Hair
General Editor
September 1993

Acknowledgments

Any project of such magnitude as a 50-year history of an organization requires the work of many people to bring it to completion. The following individuals and institutions deserve special thanks and appreciation from the Engineering Technology Centennial Committee:

- *For providing archival materials and/or photographs:* Lorretta Bailey, public relations librarian, College of Aeronautics; Frederick J. Berger, Tau Alpha Pi executive director; Maryjo Donnelly, McGraw-Hill Book Company; Fred Emshousen, Purdue University; Mary Ellen Flaherty, archivist/reference librarian, Wentworth Institute of Technology; Carole E. Goodson, University of Houston; Frank A. Gourley, Jr., division director, Engineering Technology/Industrial Technology Division, West Virginia Institute of Technology; Wayne R. Hager, Commonwealth Campus, Penn State; Brian C. Kenny, Boston University; Kent Peterson, Director, Media Services, Milwaukee School of Engineering; Gordon Rockmaker and Walter Shaw, McGraw-Hill Book Company; Robert J. Wear, retired, Academy of Aeronautics; Purdue University Programs at Kokomo; Southern College of Technology; Oregon Institute of Technology.

- *For interviewing the pioneers of engineering technology education:* Jack Beasley, administrator, Purdue University Programs at Anderson; Joseph DiGregorio, associate dean, Pennsylvania State University; Richard P. D'Onofrio, president, Franklin Institute of Boston; Carole E. Goodson, chair, Department of Civil, Mechanical and Related Technologies, University of Houston; Michael T. O'Hair, administrator, Purdue University Programs at Kokomo; William D. Rezak, dean, School of Technology, Southern College of Technology; Robert J. Wear, retired, Academy of Aeronautics; Lawrence J. Wolf, president, Oregon Institute of Technology.

- *For secretarial assistance:* Sarah Morris, West Virginia Institute of Technology; Sara Schwake, Oregon Institute of Technology.

- *For reviewing "The Presence of the Past":* Rolf Davey, Richard Ungrodt, Robert J. Wear, Ernest Weidhaas.

- *For design elements:* Frank Stanton, St. Louis, Missouri.

- *For computer assistance with merge of programs:* Jerry Adolph, Computerland, Klamath Falls, Oregon; Mary Hyde Martin, Mary Hyde Martin Design, Klamath Falls, Oregon.

- *For final proofing:* Maggie Wood, Klamath Falls, Oregon.

- *For assistance with the Milestones:* Carl Wolf, Austin, Texas.

- *For assistance with the index:* Jennifer Doudna, Klamath Falls, Oregon.

- *For reproduction of photos:* Mary Smothers, Oregon Institute of Technology.

- *Special thanks to:* Glencoe/McGraw-Hill for their very generous assistance with this publication and their 44 years of support of excellence in engineering technology education.

And, finally, thanks to all of the many friends in engineering and engineering technology education who have helped to make the history of engineering technology education in ASEE personal and positive.

List of Abbreviations

ABET	Accreditation Board for Engineering and Technology
ASEE	American Society for Engineering Education
ASCET	American Society of Certified Engineers Technicians
CIEC	College Industry Education Conference
ECPD	Engineers' Council for Professional Development
ETC	Engineering Technology Council
ETCC	Engineering Technology College Council
ETD	Engineering Technology Division
ETLI	Engineering Technology Leadership Institute
ICET	Institute for the Certification of Engineering Technicians
JET	*Journal of Engineering Technology*
NFET	National Forum for Engineering Technologists
NICET	National Institute for the Certification of Engineering Technologists
NSPE	National Society for Professional Engineers
REETS	Review of Engineering and Engineering Technology Studies
SPEE	Society for the Promotion of Engineering Education
TAC	Technology Accreditation Commission
TCC	Technical College Council
TEN	*Technical Education News*
TIAC	Technical Institute Administrative Council
TIC	Technical Institute Council
TID	Technical Institute Division

Introduction

by
G. William Troxler
Capitol College

The natural tendency in preparing a history of engineering technology, or for that matter any history, is to identify and celebrate the founding or seminal event. Histories seem more satisfying when presented as deep tap roots recounting the logical development of an individual, a nation, a college, or even an educational form. If it is possible to qualify as a primogenitor either in the form of an event or an individual, a critical analysis generally shows that what has been discovered is not the agent of conception but an important benchmark of ongoing progress.

So it is with engineering technology. The history is that of a movement rich in significant benchmarks, colored by flamboyant and dedicated leaders, shaped by national workforce needs, constrained by ever-changing politics. Those who read this book with the hope of learning the founding date and the names of the founding agents will be disappointed.

The history is best read as Baroque tapestry is viewed. These beautiful works of art were created by many unnamed artisans acting together. They traded ideas, challenged one another's creativity, worked both cooperatively and competitively. They took years to weave individual threads and produce moving representations of something meaningful to their times. This analogy is effective and accurate save one important detail. The flaw is that a tapestry is static and engineering technology education is not. At the very least, engineering technology's tapestry will never be finished.

This history presents the artisans of engineering technology through wonderfully human oral histories. Many of the pioneers of engineering technology education tell their personal stories of the early years in vivid detail. They speak with affection about the threads they and their institutions wove into engineering technology's unfinished tapestry.

The detail of engineering technology's history lies in the chapters explaining the Engineering Technology Division and Council, the *Journal of Engineering Technology*, the development of the Engineering Technology Leadership Institute, and the character sketches of McGraw Award recipients. Here the nuts and bolts of history are sorted, cataloged, and displayed. The combination of the collective memory of the pioneers and of engineering technology's history as recorded in its publications, reports, and the minutes of its meetings displays an extraordinary consistency.

Critics might say that engineering technology never seems to get beyond a few issues. But those of us who love this field will say that we have always been true to the issues which define it. The earliest records and memories of engineering technology education show concerns, debates, and reports about terminology, curriculum, accreditation standards, faculty credentials, laboratory equipment, recognition of graduates, and the need for and status of certification for engineering technology graduates. In the early years, the discussion about credentials was over certificates. Today, engineering technology educators debate curricula leading to master's and doctoral degrees.

Several themes within the history may help make a whole cloth of the individual threads. One is the perennial self-study and critical review of engineering technology education by engineering technology professionals and others. Several authors make reference to various studies of technical education. The first was the 1931 Wickenden and Spahr Report, and the most recent is the 1990 ASEE Spectrum Report. It is a great tribute to engineering technology educators that they have a record of 60 years of dissatisfaction with the status quo and of aspirations for greater quality and impact. Some engineering technology professionals may look back at the period from 1962 through the present with weary eyes and wonder about excess. During a large portion of those three decades, a major report on engineering technology was issued nearly every 18 months. By 1977, reports were so numerous that ASEE even issued a report on the reports entitled, "Review of Engineering and Engineering Technology Studies" (REETS). The history shows engineering technology education to be a restless quest for excellence.

A second major trend in the history is the constant concern with establishing the field as a professional-level discipline. Many of the oral histories and much of the written record describe the struggle with professional registration for the engineering technology graduates. The creation of the National Institute for the Certification of Engineering Technologists was one effort intended to enhance the professional status of engineering technology. Some of the oral histories describe the difficulty in securing membership for engineering technology graduates in the professional societies. The ongoing struggle with the federal Civil Service Commission to gain recognition of bachelor-level engineering technology graduates is an example of the persistence of engineering technology educators in dealing with issues of professional status for graduates.

Perhaps the most significant and successful example of engineering technology education's commitment to professionalism is the creation and nurturing of the Tau Alpha Pi Honor Society for Engineering Technologies. On the campus of Southern College of Technology in 1953, Tau Alpha Pi evolved from an honors club. From 1953 to 1973, 17 additional chapters were established on other campuses. In 1973, Dr. Frederick J. Berger became the executive director and founded Tau Alpha Pi Honor Society for Engineering Technologies in its present decentralized structure.

Tau Alpha Pi is to engineering technology what Phi Beta Kappa is to the liberal arts and Tau Beta Pi is to engineering science. The society recognizes and honors students at both the associate and baccalaureate levels. Eligibility for membership requires high scholastic achievement, nobility of character, and qualities of leadership. Since 1973, the society has grown to over 135 chapters across the nation and has become an integral part of the infrastructure of engineering technology education.

A third major theme and the subject of two chapters deals with the professionalism of engineering technology educators. In 1976, the first Engineering Technology Leadership Institute was held at Indiana State University at Evansville. The conference was the result of Tony Tilmans' and Sam Pritchett's vision to expand to a national scale the annual meeting of the three-year-old Mid-Mississippi Valley Technology Program Directors.

ETLI's purpose was to assist in the development of leaders within the engineering technology community. ETLI achieved this purpose by providing an open forum for discussions of problem areas pertinent and current to engineering technology. Much of the cohesiveness and goodwill within this community can be attributed to ETLI's unifying and cathartic influences. ETLI certainly influenced and enhanced the entire engineering technology leadership cadre over a 15-year period.

Another major contributor to the theme of increasing professionalism for engineering technology educators was the creation of the *Journal of Engineering Technology*. *JET*'s origins lie in the 1981 ETLI meeting, during which a small group discussed creating a journal devoted to the continuing national advancement of the field of engineering technology. Under the chairmanship of Larry Wolf, members Michael O'Hair, Ron Scott, Durward Huffman, and Ken Merkel worked to produce the inaugural issue of *JET* in March 1984. In the spring of 1994, *JET* celebrated a decade of service to the engineering technology community. *JET* has been a credit and an aid to all engineering technology educators. It has served as a publication vehicle for faculty and a central communications point for the engineering technology community.

This history is focused exclusively on engineering technology education. A few points on how the field fits within the histories of education and engineering education may help the reader integrate engineering technology within a broad and long view of history. Five connections seem helpful.

First, all of engineering education can be seen as a derivative or copy of the *École Polytechnique* of Paris, France. Founded in 1794, it was the first school dedicated to preparing people to assume the professional responsibilities of engineers. The first non-military school of engineering in the U.S. was Rensselaer Polytechnic Institute, founded in 1824.

The second connection must pay tribute to nineteenth-century America's leap into the industrial revolution and the need for technical workers. In many cities, numerous technical institutes were established for the purpose of educating the engineering technologists of the day. Schools now famous, such as Massachusetts Institute of Technology; schools now merged into others, such as Ohio Mechanics Institute; schools now gone, such as Spring Garden College; schools now no longer technical, such as The Maryland Institute; and schools still true to their founding missions, such as Cogswell Polytechnical College, prepared five to seven generations of engineering technologists before it seemed essential to distinguish graduates of such programs with unique titles and degrees.

The third historical connection with engineering technology spans the time from the Civil War and into the early twentieth century. During that 40-year period, the U.S. Congress passed three versions of the Morrill Act which created the land grant colleges. The educational institutions founded and funded by the Morrill Acts were completely committed to the agricultural and mechanical arts of the day. These mechanics programs were engineering technology in spirit. Today, many four-year engineering technology programs, such as those at Purdue University and Texas A & M, are housed within the land grant colleges created by the Morrill Acts.

The fourth element of history connecting engineering technology to larger trends in the U.S. is the distance learning movement. In 1882, the University of Chicago began awarding the equivalent of college credit for correspondence lessons. During the late nineteenth and for half of the twentieth century, many students studied college-level courses via the U.S. mail. Today, the four-year engineering technology programs at institutions including Capitol College and DeVry Institute can cite correspondence education as part of their pedigree. Other colleges and universities have extended their influence by using satellite and video technology to export engineering technology programs regionally and nationally.

The final important connection is the development of community colleges in the U.S. Beginning in the 50's and continuing into the 70's, states and counties established community colleges all across the U.S. Today, virtually all of the two-year engineering technology enrollment resides in community colleges.

The history contained within this book is a moving tribute to faculty, administrators, students, and graduates who collectively form the engineering technology community. A part of a much larger history, a record of achievement, a catalog of many things unfinished, a transcription of ongoing debates within the discipline, a litany of leadership — all of this and more is the history of engineering technology. Underpinning the history is an unspoken confidence of engineering technology educators about their work and the impact of their graduates on our society.

If she had known about engineering technology educators, it is likely that Eleanor Roosevelt would have welcomed the application of her words to the people of this history: "The future belongs to those who believe in the beauty of their dreams."

Chapter 1

The presence of the past:
A Brief History of the Engineering Technology Division and the Engineering Technology Council

by
Frank A. Gourley, Jr.
West Virginia Institute of Technology

In a Few Minutes

The first thing that strikes one in reading back through the meeting minutes of engineering technology educators in the American Society for Engineering Education is the quality, leadership, and dedication of the people involved in the engineering technology education

 See Appendix I for photo captions and credits.

movement in the United States.[1] There were long, well-attended business meetings of the division; minutes of annual meetings sometimes exceeded 30 pages. Numerous committees functioned and reported on a regular basis. And, in 1951, there was a "Manual for Officers and Members of the Technical Institute Division."

Committees dealt with issues such as defining technical education, establishing criteria for quality programs, identifying and preparing faculty, and determining what categories of programs should be included in the focus of the organization (at one point, the question was even raised as to whether agriculture, business, and health technology programs should be included in the organization). They faced decisions about what criteria programs should meet in being evaluated by an accrediting agency, and what to call the degrees and the programs being offered.

During those early years, there was a close relationship with the Accreditation Board for Engineering and Technology, then called the Engineers' Council for Professional Development. One book company was significantly involved in the ongoing function of the division. *Technical Education News*, published several times a year by the McGraw-Hill Book Company, regularly reported on activities of the division and was widely disseminated to technical schools and other agencies involved in technical education.

There was also a continual effort to recruit new members and involve them in the division, particularly during the early 1960's, when community colleges were being formed and engineering technology programs were being established in these institutions. This organization of engineering technology educators, primarily formed from private and proprietary schools up to that point, was now faced with opening its arms to a much broader community and making way for new leadership. This entire period is marked by a sense of dynamic involvement of many people dealing with a challenging field. Even in the 1950's, attendance at some of the business meetings exceeded 100.

The struggle for recognition and respect among peer groups of educators has been an underlying theme for some of the actions taken by this organization, but efforts to promote leadership development and quality education among engineering technology educators

---

[1] See Chapter 5 for names and institutional affiliations of individuals and information on specific activities in which they were involved.

nationwide have been the primary motivating forces over the years. These efforts have taken the form of projects, committee activities, individual service, leadership development, focuses on issues, affiliation and articulation with other agencies, research and publications, reporting, and service as a professional support group.

Division Development and Council Conception

> *Technical [e]ducation is surveyed on a national basis and it is found that the engineering college is attempting to cover too large a field; part of what it is attempting could be much better done by a different type of school with briefer and more intensive training — The Technical Institute.*[2]

The role of engineering technology education in the society and the activities of those involved in this type of education were gradually formulated and disseminated. The establishment of the engineering technology component within ASEE is due to the efforts of a consortium of representatives, from a number of technical institute-type schools, who banded together in the early 1940's to meet during annual conferences of the society. When the Society for the Promotion of Engineering Education was reorganized in 1946 as the American Society for Engineering Education, technical institute education was given a coordinate place with other groups by being granted the division privilege of having a representative on the executive committee. At this time, it also established a sort of "super committee" composed of representatives of different institutions.

Committee of 21

The Committee of 21 provided leadership for the overall operation of the Technical Institute Division between 1946 and 1962. These individuals were dedicated to the cause of technical education and

---

[2]C.F. Scott,"Final Report of the Chairman of the Board of Investigation and Coordination," *SPEE Proceedings* 42 (1934): 108.

engineering technology education. The group was comprised of representatives of proprietary, private, and public technical institutes.

The Committee of 21 was, in effect, what is now the Engineering Technology Council. It provided institutional input into the overall operation of the organization. Each year, seven individuals were appointed for a three-year term of office. Officers for the division were elected from this group. At the annual meeting, the division's sessions were well attended, and the papers and discussions furthered the professional development of engineering technology educators.

TID/TIC

When ASEE underwent another reorganization in 1962 to strengthen the responsibilities of elected representatives, the place of engineering technology education was further recognized. The Technical Institute Division continued as a resource for individual engineering technology educators who were members of ASEE. A new category of institutional membership for technical institutes was established, with ECPD accreditation required for eligibility. Representatives of institutional members constituted the Technical Institute Council, with its chairman being a vice-president of the society and a member of the board of directors. The Committee of 21 was then phased out.

Reflecting updated nomenclature, the TIC was later renamed the Technical Institute Administrative Council in 1965; then the Technical College Council in 1971; the Engineering Technology College Council in 1981; and, finally, in 1987, the Engineering Technology Council (see Appendix A for a timeline of name changes). The place of the division within ASEE has been stable since 1962, the only change in status being the name change to the Engineering Technology Division, made in 1971. Most of the committees which had been part of the division over the years were moved under the council.

Projects in Perspective

> *Incidentally, we are far from happy at the preparation of the average high school graduate in such basic subjects as English, mathematics, reading, and spelling. I wonder if it would not be a worth-while* [sic] *project of the American Society for Engineering Education to encourage the high schools to strengthen these subjects, even at the expense of the highly controversial extracurricular activities, so that when a boy enters our institutions, we won't have to waste so much time teaching him the things he should have learned in high school.*[3]

The engineering technology educators involved with ASEE have undertaken a number of projects over the years that have had an impact on engineering technology education nationally.

In the 1940's, the linkage of the Technical Institute Division with the *Technical Education News* provided a platform for communicating nationally regarding developments in the field. This was one of the first efforts to provide national exposure for engineering technology education. The first issue of *TEN* described the division's formation (see Appendix B for a reprint of the original article). The division considered merging with other interest groups, such as the Junior College Division.

Attempts to establish quality standards began as early as 1949 (and probably before) with the work of the Curriculum Development Committee, which issued at least seven reports analyzing the content of accredited courses. The establishment of the National Committee of 21 provided an organizational structure for the division to face key issues in the 1950's in the field of "technical institute-type" education.

In the 1950's, a number of division committees focused attention on the resolution of issues, such as promoting national visibility, terminology, teacher training, the place of general studies in the programs, placement of graduates, curriculum surveys, curriculum development, surveys of enrollments and graduates, and relations with professional societies for technical institute-type and engineering technology education.

---

[3] C.S. Jones, *Technical Education News* (1954).

Annual surveys of engineering technology enrollments and graduates began in 1953 and were conducted until the Engineering Manpower Commission took over this function in 1978. Early surveys included full- and part-time students in YMCA's, industry, junior colleges, technical institutes, college and university programs.

In 1953, "The Engineering Technician" pamphlet was printed to inform prospective students of careers in engineering technology and provide specific information on a variety of program areas. Over 600,000 copies were distributed in approximately 15 years.

The Committee on the Place of General Studies in the Technical Institute Program conducted a survey in 1955 to determine the extent of general studies courses in technology programs. The "National Survey of Technical Institute Education," supported by the Carnegie Foundation, was completed in 1959. In the 1960's, engineering technology educators sought to establish their position in the organization. A study to establish criteria for quality programs, the "Characteristics of Excellence in Engineering Technology Education" report, was completed in 1962, and the division initiated an effort to write a council and division history. Section representatives were identified to promote engineering technology sessions at ASEE section meetings. A teacher exchange program was functioning. There was much attention given to the four-year engineering technology programs. With growing public support of technical education, there was significant growth of membership and the subsequent challenges of assimilating new faces compounded by "threats of extinction" to many of those who had provided key leadership in the movement over the years.

During the 1970's, two additional national annual meetings were established for engineering technology educators: one to help promote industry involvement and the other to develop leadership capabilities. The "Goals of Engineering Technology Education" study, supported by the National Science Foundation, was completed in 1972. The society also had to adjust to the rapid increase of engineering technology members. The functioning of section representatives was re-established. The efforts to complete a history of the council and division ended with the historian's death and the subsequent loss of materials that had been collected for this purpose. Publication efforts resulted in establishing an annual May issue of *Engineering Education* devoted to engineering technology education and the publication of several resource monographs.

The 1980's saw an increase in activities: the mini-grant program was established, a joint committee of ETD/ETC/ETLI identified goals and objectives, the *Journal of Engineering Technology* began, operational guidelines were written for the division and the council, the national archives for engineering technology were established, a number of additional monographs were produced, and special interest groups began meeting at the annual conference.

ETD Mini-Grants

The ETD "mini-grant" program supported projects considered to be beneficial to division membership. These projects were intended to be national in scope and broad-based, but conducted by someone at an institutional level.

Beginning in January 1980, projects have been approved on an ongoing basis. These projects were initially funded at $100 and were increased to $250 in 1985.

Mini-Grant Projects
1980-1992

1980

"A Survey of Administrator Development Needs." Issac A. Morgulis, Ryerson Polytechnic Institute, Toronto, Ontario, Canada.

"Engineering Technology" (brochure). Harris Travis, Southern Technical Institute, Marietta, Georgia.

"National Listing of Institutions Offering Engineering Technology." James Todd, Vermont Technical College, Randolph Center, Vermont.

"Reference Materials for the Technical Library." James Weatherly, Murray State University, Murray, Kentucky, and Frank Gourley, North Carolina Department of Community Colleges, Raleigh, North Carolina.

1981

"Careers in Engineering Technology" (brochure). Harris Travis, Southern Technical Institute, Marietta, Georgia. [Funds provided for printing.]

"Survey of Non-Renewal and Non-Members (of ETD)." James Forman, Rochester Institute of Technology, Rochester, New York.

1982

"Survey of Graduate BET Students." Harris Travis, Southern Technical Institute, Marietta, Georgia.

"Women in Engineering Technology Programs." Diane Rudnick, Suffolk University, Boston, Massachusetts.

1983

"Current Status of Accredited Baccalaureate Programs in Electronic Engineering Technology." Lyle McCurdy, California State Polytechnic University, Pomona, California. [A monograph, with T. Kanneman as co-author, was published in 1986 as a report of this study: "Characteristics of TAC/ABET Accredited Baccalaureate Programs in Electronic Engineering Technology: Organization, Curricula and Objectives."]

1984

"Survey of Competency Requirements of Engineering Technology Administrators." Frank Gourley, Carolina Power & Light Company, Raleigh, North Carolina. [Funds provided for printing.]

1986

"Technological Interface Media for Education." David W. Brown, Kansas Technical Institute, Salina, Kansas.

1987

"Computer Software for Engineering Technology." Robert J. Buczynski, Penn State – Berks Campus, Allentown, Pennsylvania.

1989

"Group Technology and Its Influence in the Educational Community." Dan Sharp, University of Dayton, Dayton, Ohio.

1990

"Survey of Instruction in Electromagnetic Fields and Waves in Electrical Engineering Technology Curricula." Steve Walk, University of Maine, Orono, Maine.

1991

"Survey of Audio Visual Materials to Support Technical Specialty Areas." Robert Herrick, Purdue University, West Lafayette, Indiana.

1992

"Analysis of Mechanical Design Engineering Technology and Related Programs to Support Curriculum Development." Michael R. Kozak, University of North Texas, Denton, Texas.

The Professional Development Recognition Program

In the fall of 1982, after an ABET accreditation visit, a couple of ETD members discussed the need to stimulate professional development among engineering technology faculty and institutions. As a result of this discussion, they searched to identify a model at the national level that might serve as an incentive to promote professional development. They located one a few months later, a state professional engineering society which had such a program in place.

In 1983, a project began to develop a similar professional development recognition program for engineering technology educators. However, preliminary work to develop criteria and standards was put on hold in 1985, due to questions about the program's viability and implementation procedures.

In 1987, ETD re-initiated this program, completed criteria, and developed guidelines designed to recognize professional development in four basic areas:

- Professional work experience

- Participation in professional organizations

- Professional activities

- Educational/development activities

As conceived, interested individuals would complete forms and participate for three years, the period designated for certification, with a renewal application required at that time to remain certified. Such a program would, supporters felt, stimulate engineering technology educators and focus their professional development so as to promote ongoing activities. Some of the larger institutions in the country supported it, and in 1989 ASEE headquarters approved implementation on a trial basis for three years. To date, however, the program has not been implemented. Such is the fate of some projects in volunteer organizations.

ETD Special Interest Groups

In 1984, special interest groups (SIG's) were initiated after a membership survey determined interest in special topics. SIG's held evening meetings at the annual conference, to promote communications among members, and focused on topics such as computer-aided design, telecommunications, robotics, automated manufacturing, and surveying/construction.

Meeting minutes and lists of attendees were kept. The SIG's met with reasonable success for several years and provided a way to address new and emerging technologies, as well as sub-discipline interests.

In addition, attempts were made to identify the discipline interests of all ETD members, via a membership directory, but these efforts have not been successful to date.

Section Representatives

In the mid-1960's, section representatives were identified to work with institutional representatives, promoting engineering technology education activities at the section level within ASEE. One representative was appointed from each of the 12 geographic sections. Their initial role was to promote geographic section programs for engineering technology educators, an effort stimulated by a national section activities coordinator.

During their regular section meetings, they conducted specialized sessions for those with engineering technology interests. Although this function was discontinued in 1971, seven years later the concept was re-introduced, this time as an engineering technology newsletter committee. Once more, 12 section representatives were appointed. Two years later, the responsibilities broadened and a section activities coordinator was again appointed.

Section representatives were encouraged to engage in a number of activities to promote the division on a local level:

- Contact individual institutions and request news items for the ETD newsletter
- Suggest possible articles for publication in *Engineering Education*
- Identify topics of interest that might be presented at section or national meetings
- Promote the participation of members in geographic sections by encouraging ETD-sponsored sessions at meetings
- Recruit new members
- Communicate information about the activities of the ETD.

The section activity coordinator's role was to promote the division on a regional level:

- Maintain contact with the section representatives
- Work with the section representatives to provide sessions on engineering technology at section meetings
- Facilitate communications among section representatives regarding successful section activities
- Maintain the liaison with section representatives and the newsletter editor
- Recruit section representatives as necessary.

The section representatives and section activity coordinator continue to function with these same charges.

Program Directory

Another major project has been the *Directory of Engineering Technology Institutions and Programs*. This project was first conceived in the early 1970's, when efforts were made on a state-by-state basis to collect information on engineering technology programs and the institutions offering those programs. However, efforts to coordinate a committee of 50 to prepare the information just did not gel. Later, a mini-grant supported this effort. The result was a listing of about 776 institutions which offered engineering technology programs, but it did not identify the specific programs or provide contact information for the institutions.

It was not until 1989 that this project moved towards fruition. In 1990, using Engineering Manpower Commission data, the first directory was completed, listing over 400 institutions which offered a total of more than 1,700 engineering technology programs. This directory provided contact information on the institution and identified the programs offered, as well as indicating the level (A.S., B.S., M.S.) and accreditation status of each program.

At Issue Here

I also remembered a sentence in one of Dean Hammond's sketches of the [s]ociety history when the [s]ociety was 40 years old: "The chief impression one obtains from Volumes I and II of the Proceedings is the unchangeableness of the fundamental aims and ideals of engineering education. One finds the same expressions then that we use today and perhaps consider them to be original and new. One finds the same problems discussed with which we are now confronted.[4]

The division and council have dealt with a number of issues over the years, including the relationship of engineering technology to engineering and industrial technology, the role of baccalaureate and graduate engineering technology programs, the relationship between the ETD and the ETC, involvement with international technical education, the division membership representation in society leadership, industrial involvement, relationships among the three "P's" (public, private, and proprietary institutions), and division and council finances.

Major Issues

Questions regarding the relationship between engineering technology, engineering, and industrial technology have been addressed in various formal and informal reports and in individual articles and presentations. The 1982 ETC By-Laws Committee report presented the need to involve institutions offering non-accredited engineering programs in the organization. The relationship between the ETD and the ETC was clarified through work of the Program of Work Committee in 1981-83 and through subsequent division and council actions.

Other reports to clarify issues of consistency and quality in engineering technology education include the "Characteristics of

---

[4]Nathan W. Dougherty, "Foundation for Our Future, 75 Years of Progress: American Society for Engineering Education," *Engineering Education* 58, no. 9 (1968): 1029.

Excellence in Engineering Technology Education" report, published in 1962, and the 1972 "Engineering Technology Education Study." Other issues have been addressed through committee action, resolutions, white papers, motions, or group interchanges. For example, in the early 1980's, when engineering technology educators felt that they did not have sufficient representation in the overall society, the ASEE Board of Directors was approached with the concept of rotating society presidents among councils. While this proposal was not approved, an indirect result was the unopposed running of Richard Ungrodt, a key engineering technology educator, for ASEE president a couple of years later.

Over the years ETD has expressed interest and concern in involving industry in the overall activities of ASEE related to engineering technology education. This has been reflected through the involvement of several industry personnel in leadership positions of the division and regular efforts to recruit members from industry to interact with the division.

College Industry Education Conference

Another effort to involve industry was a new winter meeting. Although ETD winter meetings date back to the 1950's, when a gathering was typically held in October, these meetings were phased out in the early 1960's. But by 1976, the division began what became known as the College Industry Education Conference, co-sponsored by three other divisions: Cooperative Education, Continuing Engineering Studies (now Continuing Professional Development), and Relations with Industry (now College-Industry Partnerships). Now a part of division tradition, CIEC has provided both a forum for interaction with industry and a support for the division's activities in other ways, including providing a financial base used to underwrite a number of projects.

The idea for the CIEC was formulated by the Relations with Industry Division in 1974 and presented to the other divisions, which quickly approved it. The conference was designed to promote "industry-education dialogue and communication." Requirements were that it attract industrial participation, draw 500 registrants, be held in mid-academic year, include four joint sessions, have no dominating division, have at least one nationally known speaker, run concurrently with a meeting of the deans of engineering, and provide participation of practicing engineers. Meetings were to be held in the southern tier of the United States and to alternate geographically. The CIEC has generally followed these guidelines.

CIEC's success has resulted in an ongoing opportunity for engineering technology educators to interact with industry personnel about topics related to industry cooperation and interaction. It also has allowed the division to become financially solvent, enabling it to support other projects such as the *Journal of Engineering Technology*, the mini-grants, and the publication of monographs. CIEC offers a formal program that runs for three full days, with a day of business meetings and workshops preceding the conference.[5]

College Industry Education Conference
1976-1994

		ETD Program Chair	Conference Chair
1976	Orlando	Gerald A. Rath	Joseph M. Biedenbach
1977	San Antonio	Issac I. Morgulis	Carl P. Houston
1978	San Diego	Frank V. Cannon	James P. Todd
1979	Tampa	James W. Bannerman	Frank T. Carroll
1980	Tucson	Anthony L. Tilmans	Joseph M. Biedenbach
1981	Lake Buena Vista	John D. Antrim	Martha J. Johnson
1982	San Diego	Ronald C. Paré	Joseph C. Mehrhoff
1983	Lake Buena Vista	Harris T. Travis	Glenn A. Burdick
1984	Dallas	Kenneth K. Gowdy	Bill L. Cooper
1985	San Diego	Samuel L. Pritchett	Luther Epting
1986	New Orleans	Durward R. Huffman	Kenneth K. Gowdy
1987	Lake Buena Vista	William D. Rezak	Robert W. Ellis
1988	San Diego	James A. Hales	Joe S. Greenburg
1989	New Orleans	Lyle McCurdy	R. Neal Houze

---

[5]For more information, see Frank Burris, "Origins and Traditions of CIEC," *College Industry Education Conference Proceedings* (1993): 4-9.

		ETD Program Chair	Conference Chair
1990	Lake Buena Vista	William S. Byers	James A. Hales
1991	San Diego	W. David Baker	William Wilhelm
1992	Las Vegas	Larry D. Hoffman	Anthony L. Rigas
1993	Lake Buena Vista	Robert English	Gary L. Hamme
1994	San Antonio	J. Dale Pounds	Richard M. Moore

Committed Committees

From the 1940's through the early 1960's, the Technical Institute Division had a strong committee structure at work. These committees focused on completion credentials, cooperation with the Office of National Defense, curriculum development, manpower studies, membership, program, relations with industry, student selection and guidance, technical institute studies, and teacher training. In the early 60's, with the reorganization of ASEE and the separation of ETD and ETC, many of the Technical Institute Division committees became ETC committees. Since that time, most standing committees are part of the Engineering Technology Council.

In 1972, the council had 15 standing committees and one *ad hoc* committee. Some of those committees have been phased out and others have changed names. In 1992, the council had 12 standing committees and one *ad hoc* committee. To support the activities of these committees, ETC has had regular meetings where the committees reported on activities and outcomes of their efforts. Involvement on committees has been one way of integrating new people into the organization and of providing recognition to those promoting the goals and activities of engineering technology education.

The Program of Work Committee

The Program of Work Committee was established in February 1981 to identify goals, objectives, and activities of the Engineering Technology Division. The goals of the division were developed to parallel the goals of ASEE. Objectives and activities were identified to enhance these goals. Early in its deliberations, this six-member committee realized that determining these items also involved the Technical College Council and the Engineering Technology Leadership Institute. The result was an expansion of this effort to a 13-person committee composed of representatives from all three groups. They met in June 1981 and again in February and June 1982 to clarify the relationships and responsibilities of the three groups and develop goals, objectives, and activities for each. Another outcome of these meetings was that the ETLI was moved, organizationally, under the Technical College Council.

Once there was agreement among the three groups, program of work schedules were developed based on the objectives and activities which identified responsibilities of each officer and committee of the ETD. In 1983, the program of work schedules were combined with officer responsibility and committee function statements in one document, "The Guidelines of the Engineering Technology Division." The Program of Work Committee dissolved in 1984, having completed its charge. In 1987, the ETD Goals and Activities Committee updated the "Guidelines." These were the first written guidelines prepared since the 1951 "Manual for Officers and Members of the Technical Institute Division."

Goals and Activities Committee

In 1985, the Goals and Activities Committee was established to stimulate ETD activities by identifying potential projects and project leaders to involve more of the membership in division activities. Some of this committee's projects have included updating the division guidelines, initiating special interest groups at the annual conference, and establishing an engineering technology educators' professional development recognition program. However, the underlying thrust has been on strategic, long-range planning for the division. The committee developed a mission, goals, and objectives statement in the late 1980's which provides a continuing opportunity for discussion of where the division is going and how it wants to get there.

Lead On

Leadership development has been an important part of the activities of the division and the council over the years. The annual conference is a prime example of this effort.

From the early years, when all attendees sat around in the same room for a couple of days discussing and deciding on the issues, to the present, when three or four sessions on different topics run concurrently, the primary purpose is still leadership development. The annual conference program has provided a place where a variety of discipline interests could be addressed, such as electrical, mechanical, civil, administration, department heads, special interests, and minorities and women. Presenters, moderators, chairs, members of committees, program coordinators, and members of the audience — all have been involved in developing their leadership capabilities through the progression of attendance, involvement, development, and promotion. See Appendix C for listings of ETD and ETC officers.

Other meetings that have developed leadership capabilities include the annual CIEC and programs at section meetings. The Engineering Technology Leadership Institute was established in the mid-70's to provide a specific outlet for developing leadership potential. This effort has served its purpose well and has been an outreach program to new engineering technology educators not directly involved in the national functioning of the division. Chapter 2 details ETLI's development.

A Little R and R (Researching and Reporting)

The division and council have also been involved in researching and reporting over the years. Early studies of engineering education usually included comments, references, or full sections on engineering technology. The first definitive study of engineering technology education was conducted under the auspices of the Society for the Promotion of Engineering Education and published in 1931 as the Wickenden-Spahr Report, which specifically identified the engineering technology sector as an important part of the engineering community.

The 1955 Grinter Report is considered by many to be instrumental in the development of four-year engineering technology programs.

The preliminary report proposed "professional-general" and "professional-scientific"engineering programs. The final report deleted all references to "professional-general" programs. However, the continuing needs of industry for the "professional-general" graduate led many colleges, in later years, to add two more years to existing offerings to provide baccalaureate programs.

The Technical Institute in America by Ross Henninger was published in 1959 by McGraw-Hill. It reported the results of an extensive survey of the status of technical institute education and was supported by the Carnegie Foundation.

At the request of the Engineers' Council for Professional Development, another study was conducted in 1962, titled "Characteristics of Excellence in Engineering Technology Education" and known as the McGraw Report. This report was the basis for upgrading the accreditation criteria for associate degree programs.

In 1972, the ASEE "Goals" study was completed and titled "Engineering Technology Education Study." This report was a key document in developing differential criteria for the accreditation of associate and baccalaureate degrees.

In 1977, ASEE released the "Review of the Engineering and Engineering Technology Studies" (REETS). Conducted as a review of existing studies to determine the need for another major study of the profession, there was no significant impact on engineering technology education.

The "Quality in Engineering Education Project" (QEEP) report was completed in 1986. The report used the generic term "engineering" throughout to include both engineering and engineering technology. To date, there has been no significant impact on engineering technology education from this report.

Other studies have been the Engineering Technology Development Committee work in the early 80's, the Engineering Technology Council By-Laws Committee effort in 1982-83, mini-grant projects beginning in 1980, and Research Committee activities of the 80's.

Beginning with *Technical Education News*, which reported on meetings of the division in the 1940's, 1950's, and 1960's, the division clearly has been "in the news." Publication efforts since then have been spearheaded primarily by the ETC Publications Committee and the *Journal of Engineering Technology* Editorial Board. Other efforts have included the "Technology Education Comments" column in *Engineering Education*, monographs of the TIAC Library Affairs Committee, the ETD newsletter, the engineering technology compendium, and articles in conference proceedings.

"Technology Education Comments" Column

In the 1970's in *Engineering Education*, the "Technology Education Comments" column appeared regularly for about four years and addressed topics of interest and concern to engineering technology educators. Written by Jesse J. Defore, who assisted L. E. Grinter on the 1972 "Engineering Technology Education Study," the column usually ran one to two pages and examined a variety of topics related to engineering technology education. Below is an annual list of these topics:

1972

"Manufacturing Engineering Technology," February
"Career Education Concepts," March
"New Teachers in Engineering Technology Programs," May
"Selecting Professional Personnel for Technical Colleges," October
"Inaugurating Technical Programs," November
"The Preparation of Technical Teachers," December

1973

"The Engineering Technology Education Study: Assessment after One Year," January
"Engineering Technology: Today's Liberal Education," February
"Student Responsibility for Learning," March
"Recognizing Student Achievement," April
"The Profitable Use of Summer," May
"Articulation of Technology Programs: A Continuing Challenge," October
"Follow-up Studies in Engineering Technology," November
"Needed: A Comprehensive Student Data Base," December

1974

"Improving Customer Relations," January
"Developmental Programs: Preparing the Deficient Student," February
"'Readiness to Work' in Technical Curricula," March
"Stressing the Need for Safety in Technical Education," April
"Curriculum Structure and Technology Education," May
"Mathematics in the Engineering Technology Curriculum," November
"Physics in the Technology Curriculum," December.

1975

"Chemistry in the Technology Curriculum," January
"Technical Sciences in the Technology Curriculum," February
"The Strengths of Associate Degree Engineering Technology Programs," May
"Humanities and Social Sciences in the Technology Curriculum," November
"Investing in the Academic Future," December

1976

"Institutional Cooperation: A Need in Technology Education," January
"International Cooperation in Technology Education," December

1977

"Co-op Programs and the Development of International Technical Manpower," May

In the early 1980's, this column re-emerged for a brief time but was discontinued when ASEE changed the format of *Engineering Education*.

Journal Activity

Over the years, *Engineering Education* has published a number of issues that included articles on engineering technology education. Two special issues appeared in 1959 and one in November 1966. In 1967, when the need was expressed for another special issue, the Technical Institute Division proposed a procedure to help underwrite the cost of extra journal copies through advertisements funded by industry and educational institutions. Beginning in 1975 and annually through 1987, the May issue included articles on engineering technology education. Coordinated by the ETC Publications Committee, these issues provided a national platform for disseminating information on the field to the broader educational community.

Another platform for identifying research and reporting has been the annual engineering technology education bibliography, published in *Engineering Education*. Since 1986, the bibliography has provided a subject listing of published articles, books, and papers relating to engineering technology education and has served to identify a significant amount of the literature related to the field.

The *Journal of Engineering Technology* was established in 1984 and, with its advent, the engineering technology education community has had its own publication to address engineering technology education interests. Chapter 3 details *JET*'s development.

Library Affairs Committee of the TIAC

The Library Affairs Committee, formed in 1970 as a TIAC committee, focused on contributions it could make to engineering technology education from the viewpoint of the library.

The committee's first publication was "Engineering Technology, the Student, and the Learning Resources Center," published in May 1971. It was a study of the interaction of the library with post-high school, two-year engineering technology programs and was available from the Curriculum Center of Wentworth Institute.

In 1972, the committee published "Departmental Libraries for Technology Departments," available from the North Carolina Department of Community Colleges, Program Development Section. In 1973, "The Learning Laboratory: A System for Guided Studies in the Technical College" was published, also available from the North Carolina Community Colleges, Program Development Section. The 1974 "Reference Materials for the Technical Library" and "Audio Visual Materials for the Engineering Technologies" (1976) were available through ASEE headquarters.

The committee continued to function until 1976, but prepared no additional reports. Its activities were assumed by the ETC Publications Committee beginning in 1976.

ETC Publications Committee

The ETC Publications Committee was formed in the early 1960's to generate enthusiasm for writing among engineering technology educators. The committee's initial charges were to act as an informal sounding board for the managing editor of the *Journal of Engineering Education*, to recommend engineering technology articles for publication, to provide regular articles on engineering technology education for the *Journal*, and to increase the visibility of ET education through publications.

The Publications Committee's function has expanded since its involvement with *Engineering Education*, beginning in 1975. Its activities have included preparing and producing a variety of specialized monographs, encouraging engineering technology educators to write articles for other professional and trade journals, listing magazines receptive to articles on engineering technology, preparing bibliographies of materials on specialized topics, producing the annual bibliography of engineering technology education publications, developing brochures, reviewing brochures written by others, identifying reviewers for the *Annual Conference Proceedings*, and other similar activities. Listed below are selected projects undertaken by that committee over the years:

Brown, David W. "Audio Visual Materials for the Engineering Technologies." 2nd ed. October 1981. 35pp.

Buczynski, Robert J. "Computer Software for Engineering Technology: A Compilation." January 1988. 38pp.

Committee on Writing Across the Curriculum. *A Style Manual for Report Writing*. Klamath Falls, OR: Oregon Institute of Technology, 1986. 57pp.

Dyrud, Marilyn A. (comp.) "Engineering Technology Education Bibliography." In *Engineering Education*, April/May 1987, January/February 1989, September/October 1989, May/June 1990, May/June 1991. In *Journal of Engineering Technology*, Fall 1992, Fall 1993.

"Engineering Technology in the American Society for Engineering Education" (brochure). June 1988. 2pp.

Gourley, Frank A., Jr. (ed.). *Directory of Engineering Technology Institutions and Programs*. June 1990. 80pp.

Gourley, Frank A., Jr. "Job Competencies Required of Engineering Technology Administrators." May 1985. 40pp.

McCurdy, Lyle B., and Thomas A. Kanneman. "Characteristics of TAC/ABET Accredited Baccalaureate Programs in Electronic Engineering Technology: Organization, Curricula, and Objectives." May 1987. 61pp.

May issues of *Engineering Education*. 1975, 15 articles; 1976, 3 articles; 1977, 10 articles ; 1978, 5 articles; 1979, 6 articles; 1980, 7 articles; 1981, 8 articles; 1982, 7 articles; 1983, 11 articles; 1984, 10 articles; 1985, 10 articles; 1986, 6 articles; 1987, 8 articles.

Rudnick, Diane Tarmy. "Recruiting Women and Men into Engineering Technology Programs: Findings of the ETD/DEC Study." 1984. 38pp.

Weatherly, James G., and Frank A. Gourley, Jr. "Reference Materials for the Technical Library." 2nd ed. February 1982. 43pp.

Compendium

In 1990, a 10-volume compendium of previously published ET literature, titled *The Evolution of Engineering Technology in the Field of Engineering Education*, was released. The collection gathers together articles which have appeared in the *Annual Conference Proceedings*, the

Frontiers in Education Conference Proceedings, the *CIEC Proceedings*, and various sectional meeting proceedings. A three-volume addendum is currently available through Anthony L. Tilmans, Southern College of Technology.

ETD Newsletter

The division newsletter has been an important communication vehicle since its inception in 1955. Typically, the two issues published each year examine a variety of topics of interest to engineering technology educators. The newsletter editor is elected, serves for one year, and then becomes the ETD program chair for the annual conference.

The newsletter has taken a variety of formats over the years; it is currently approximately 30 pages (5 1/2" x 8 1/2" format). Appendix D includes data regarding newsletter topics.

A Little A and A (Affiliating and Articulating)

From the early days, the division and council have been affiliating and articulating with other groups. The division and the council were major forces in involving ABET in the accreditation of "technical institute-type" programs. Later, other professional-technical societies became active with ABET in engineering technology accreditation. ABET is now an independent organization of people from industry and education which accredits engineering technology and engineering programs, and ETD personnel continue to be involved in leadership roles.

The division and the council are closely affiliated. The two cooperate on the annual program, banquet, and McGraw Award. They also cooperate with other groups within ASEE. The CIEC, in particular, is an important area of interest and involvement by the division, since it provides structure to establish close relationships with industry personnel through the co-sponsoring divisions.

The National Institute for the Certification of Engineering Technologists has regular representation in ASEE activities, as does Tau Alpha Pi, the engineering technology honorary society. In addition, other professional societies have representatives who attend meetings or have been involved over the years, such as the IEEE, ASME, and the Engineering Workforce (Manpower) Commission.

Service with a Smile

It's hard to measure the effects of the ongoing activities of professional groups such as the Engineering Technology Division and the Engineering Technology Council. The thing that makes an organization like this "tick" is people.

The expertise that individual members of the division and council bring to ASEE is significant. The advice and counsel available during breaks between conference sessions, at the dinner table, and after hours are important parts of what happens at any engineering technology education function. The network established through such involvement has been a support to its membership over the years.

Reading a list of attendees at a division or council function is like reading a "Who's Who" in engineering technology education. Although it would be impossible to list all the key players involved with the two organizations over the years, it can be said that these people have made the organizations what they are. They have given unselfishly of their time and expertise to come together to work for the common good, representing the field and developing its people. Even when there have been sides to issues, people on each side have been professional in their interactions, and, once the issues have been resolved, they have gone on about the business of the organization as friends. What a legacy: quality, leadership, and dedication!

References/Additional Resources

American Society for Engineering Education. "Quality in Engineering Education Project." Washington, D.C.: ASEE, 1986.

American Society for Engineering Education. "Review of Engineering and Engineering Technology Studies." Washington, D.C.: ASEE, 1977.

Beatty, H. R. "The Development of Technical Education."*Journal of Engineering Education* 58, no. 9 (May 1968): 1063-5.

Burris, Frank E. "Origins and Traditions of CIEC." *College Industry Education Conference Proceedings* (1993): 4-9.

Defore, Jesse J. "Final Report: Inventory Conference on Engineering Technology Education." *Journal of Engineering Education* 58, no. 10 (June 1968): 1097-9.

Defore, Jesse J. *Technician Monographs: A Collection of Papers and Research Studies Related to the Associate Degree Programs in Engineering Technology*. Washington, D.C.: ASEE, 1971.

Dougherty, Nathan W. "Foundation for Our Future, 75 Years of Progress: American Society for Engineering Education." *Journal of Engineering Education* 58, no. 9 (May 1968): 1019-31.

"Education of the Engineering Technician."*Journal of Engineering Education* 57, no. 3 (November 1966): 185-232 [sixteen articles].

Engineering Education 60, no. 4 (December 1969): 295-308 [four articles].

"The Engineering Technician" (pamphlet). New York: McGraw-Hill, 1953. 22pp.

ETC Long-Range Planning Committee. "Engineering Technology Council Operating Procedures Manual." 1988. 50pp.

ETD Program of Work Committee. "Engineering Technology Division Guidelines." 1983. 44pp.

Grainey, Maurice R. *The Technical Institute*. New York: The Center for Applied Research in Education, 1964.

Grayson, Lawrence P. "A Brief History of Engineering Education." *Engineering Education* 68, no. 3 (December 1977): 246-64.

Grinter, Linton E. "The Bachelor of Technology Degree." *Journal of Engineering Education* 58, no. 10 (June 1968): 1099-1101.

Grinter, Linton E. (chair). "Final Report: Goals of Engineering Education." *Journal of Engineering Education* 58, no. 5 (January 1968): 374-446. [Also published separately as a report.]

Grinter, Linton E. "Summary of the Report of the Committee on Evaluation of Engineering Education." *Journal of Engineering Education* 46, no. 1 (September 1955): 25-60.

Grinter, Linton E., and Jesse J. Defore (chairs). "Final Report: Engineering Technology Education Study." *Engineering Education* 63, no. 4 (January 1972): 327-90. [Available as a separate report from ASEE.]

Henninger, G. Ross. *The Technical Institute in America*. New York: McGraw-Hill, 1959.

McGraw, James L. (director). "Characteristics of Excellence in Engineering Technology Education: Final Report of the Evaluation of Technical Institute Education." 1962. 46pp.

Society for the Promotion of Engineering Education. "Report of the Investigation of Engineering Education, 1923-29, Accompanied by a Supplemental Report on Technical Institutes, 1928-29." 2 vols. Washington, D.C.: SPEE, 1930 and 1934. [Wickenden Report and Wickenden-Spahr Report; available from ASEE.]

Technical Education News. New York: McGraw-Hill, 1941-1977.

"Technician Career Opportunities in Engineering Technology" (pamphlet). New York: McGraw-Hill, n.d. 22pp.

Wickenden, William E., and Robert H. Spahr. *A Study of Technical Institutes*. Lancaster, PA: SPEE, 1931.

Chapter 2

History of the Engineering Technology Leadership Institute

by
Anthony L. Tilmans
Southern College of Technology

In the winter of 1973, an informal group of engineering technology program directors from the Midwest met at Memphis State University to discuss problems and issues of common interest. This informal group, the Mid-Mississippi Valley Technology Program Directors, decided at the conclusion of this first meeting that they should meet the next year at the University of Southern Illinois. The following year, the group met at Western Kentucky University and at the conclusion of that meeting decided that Indiana State University, Evansville, would host the 1976 meeting.

 See Appendix I for photo captions and credits.

It was while driving home from WKU that Tony Tilmans (ISUE) and Sam Pritchett (Purdue) thought it appropriate to formalize the meetings as national conferences. During the planning stages for the first national meeting, the name "Engineering Technology Leadership Institute" was proposed, mainly because of the group's strong desire to assist in preparing future leaders within the engineering technology community.

Maiden Voyage

As the planning progressed, a discussion format evolved: the moderators would function as discussion leaders, first presenting all sides of an issue and then leading the discussion among the participants. Participants would be sent the program in advance and then encouraged to prepare for discussion of the various topics. The format proved very effective: at the first ETLI, a number of participants even brought transparencies and handouts for discussion purposes. Among the issues and topics were management by objective, program costs, faculty, utilization of industrial resources, and accreditation. At the business meeting, a motion calling for a second annual ETLI unanimously passed. A steering committee was appointed, consisting of Tony Tilmans, Sam Pritchett, John Antrim, Stanton Peters, Boyce Tate, Charles Callis, and Mike Bezbatchenko.

A major topic of discussion was whether ETLI should continue as a stand-alone organization or affiliate with ASEE through the Technical College Council. Among the topics suggested for the next meeting were professional registration and the differentiation between engineering, engineering technology, and the engineering technician.

All in all, the first meeting was a huge success, with 46 registrants and a number of local guests and dignitaries representing the following institutions and corporations:

- Alabama A & M University
- ALCOA
- Babcock & Wilcox
- Brigham Young University
- Delgado College
- Fairleigh Dickinson University
- Fairmont State College
- Georgia Southern College

- Indiana State University, Evansville
- Indiana University, Kokomo
- Indiana Vocational Tech College
- Kansas State College
- Lake Superior State College
- Louisiana State University
- Mankato State University
- Memphis StateUniversity
- Mississippi State University
- Nashville State Technical Institute
- New Jersey Institute of Technology
- Northern Arizona State University
- Northern Scientific
- Purdue University
- Purdue University, Calumet
- Rend Lake College
- Southern Colorado State College
- State Tech Institute at Memphis
- Texas Southern University
- Texas Tech University
- University of Akron
- University of Alabama
- University of Arkansas, Little Rock
- University of Maryland
- University of Tennessee, Martin
- Virginia Polytechnic Institute and University
- Western Kentucky University
- Whirlpool Corporation

More Planning

The steering committee met the following spring at Sam Pritchett's house at Barkley Lake, Kentucky, to plan the next program, as well as to develop proposals for discussion and adoption at the business meeting. These proposals soon became essential elements of future ETLI meetings. One of the first was the name, composition, and term of office of the steering committee. The name became "Executive Council" with a membership of seven: six elected by the ETLI membership for two-year staggered terms, and the seventh the coordinator for the next year's host institution. A second proposal described the purpose of ETLI:

> *The institute provides an open forum for discussions of problem areas pertinent and current to the engineering technology community.*

> Topics are carefully chosen by the Executive Council to permit members to broaden and/or solidify their understanding of vital issues, as well as to apply their joint expertise in solving current problems.

For voting purposes, the ETLI membership consisted of those in attendance at a respective business meeting. The business meetings were of vital importance as they were very issue-oriented and controversial. Among these issues were engineering technology's niche in the overall engineering spectrum, ETLI constitution and by-laws, the possibility of master's degree programs in engineering technology, affiliation with the American Society for Engineering Technology (Texas group), and affiliation with ASEE through ETCC.

After many years of discussion, ETLI decided to request affiliation with ASEE/ETCC, which has proven to be extremely rewarding to the organization. Since the 1984 ETLI, ETCC has held its business meeting prior to ETLI.

One of the highlights of ETLI is the now infamous Cracker Barrel Session which was, for many years, under the helm of Sam Pritchett, who kept the discussion lively and controversial. After a full agenda of sessions, a business meeting, and a luncheon, the participants are ready for a little "attitude adjustment," with liquid refreshments, fruit, crackers, and cheese. The room is set up in a circular seating arrangement with a huge "soapbox," ready for an eager presenter to speak his/her piece. Although most of the participants know each other, the presenters must identify themselves and their affiliation before speaking.

Once things get going, there is rarely a lull in the action; there always is a hot topic for discussion, as well as prodding by an ever-capable coordinator. So that participants can have dinner by a reasonable hour, the coordinator usually has to bring the session to premature closure.

Due to the success of ETLI, a Site Selection Committee now reviews host institution proposals and determines host institutions two to three years in advance. See Appendix E for a list of ETLI locations and conference chairs.

NFET and ETLI

The 1987 ETLI, held in Nashville, was unique because it served as a preparatory discussion session for the National Forum for Engineering Technology, designed to deal with major topic areas affecting the delivery of engineering technology education throughout the country. Roughly formatted after the National Congress of Engineering Education held in 1986, NFET addresses issues and areas unique to engineering technology education.

Early in 1987, ABET's Technology Accreditation Commission accepted the challenge to conduct NFET as the 1987 ABET annual meeting. As initial planning began, coordinators learned that ETLI was also being held in Nashville, beginning the weekend prior, and decided that the two events be linked, since many of the same people would attend both conferences. A steering committee was formed with Lawrence J. Wolf, TAC chairman-elect; Russel C. Jones, ABET president-elect; Anthony L. Tilmans, ETLI; and David R. Reyes-Guerra, ABET executive director.

Ray L. Sisson, University of Southern Colorado, was named program chairman for NFET, and Warren W. Worthley, Indiana University-Purdue University at Fort Wayne, was asked to chair the planning of sessions to be presented.

Five major issues were identified for debate, and an issue chairman was appointed to develop each area:

- Faculty issues (Warren W. Worthley, chair)

- Curriculum issues (Fred W. Emshousen, chair)

- Industry and professional issues (Leo Ruth, chair)

- Long-range engineering technology issues (Stephen R. Cheshier, chair)

- Engineering/engineering technology interface issues (Stephen R. Cheshier, chair)

As a final consensus to proceed, these issues were presented to representatives of ABET member societies for approval and the program finalized for the Nashville meeting.

NFET used high-technology equipment to provide participants with immediate feedback on their voting activities. The issues in each session were presented to the participants using Harvard Presentation Graphics software, an NEC portable computer, a Kodak datashow, and an overhead projector and very large screen. At the close of each session, participants voted on National Computer Systems, Inc. mark-sense sheets, which were rushed to Nashville State Technical Institute for computerized scoring. These results were recorded session by session, using the Harvard Presentation Graphics slideshow, and were presented at the closing session in bar graph form, showing the "mandates" of high acceptance and the "strong rejections" of negative responses.

Approximately 178 participants attended the forum and provided lively interaction and solid feedback through their debate and voting. At the close of each session, questions from the floor were requested in preparation for the forum's final session, yielding approximately 69 issues received from participants. Of those, 12 questions were selected and voted upon by the body in the final session. The wrap-up session showed statistical bar graphs of the voting data from each issue.

Following NFET, Russel C. Jones appointed a follow-up committee composed of Stanley M. Brodsky, Edward T. Kirkpatrick, Robert L. Reid, Anthony L. Tilmans, Lawrence J. Wolf, and Warren W. Worthley (chair), to ensure that the "mandates" and "rejections" received special attention as possible modifications of TAC/ABET criteria. The NFET Follow-up Committee met on February 12, 1988, to set the action timing for each of the NFET issues. Special *ad hoc* committees were formed within TAC to review issues with the expectation that some actions would take place prior to the TAC annual meeting in July, 1988.

New Directions

Following 17 consecutive successful institutes, planners for the 1993 meeting decided, on a trial basis, to conduct ETLI-related workshops immediately prior to the 1994 CIEC in San Antonio, Texas. The reason for this action was based primarily on reduced budgets limiting travel, with the understanding that ETLI attendees would also attend CIEC and thus attendance should be at least normal. Hopefully, a number of CIEC attendees would also register for ETLI workshops.

Such flexibility in a time of constraining budgets illustrates the versatility associated with the leadership qualities which ETLI strives to develop.

Chapter 3

Journal of Engineering Technology

by
Michael T. O'Hair
Purdue University Programs at Kokomo

and

Lawrence J. Wolf
Oregon Institute of Technology

The purpose of the Journal of Engineering Technology *is to promote and nurture engineering technology as a distinct body of knowledge, to foster inquiry in engineering technology, and to disseminate the results of such inquiry.*

The Early Development

"In late 1981, while attending the Engineering Technology Leadership Institute held at Arizona State University in Tempe,

 See Appendix I for photo captions and credits.

Arizona, a small group of [ET] educators met to discuss the possibility of publishing a new journal which would be devoted to the continuing national advancement of the field of engineering technology.[1] In November 1981, a sub-committee of the Communications Committee of ASEE's Engineering Technology Division was formed to study the type of journal that would be the most appropriate for the engineering technology community. Larry Wolf served as chair, with members Michael O'Hair, Ron Scott, and Ken Merkel. They called themselves the Journal Study Committee.

Chairman Wolf conducted the first meeting via a telephone conference on December 9, 1981. During that meeting, the committee drafted a document outlining a purpose statement, an editorial policy, personnel, a start-up procedure, and an editorial review process which would be used in meetings with appropriate committees at the 1982 College Industry Education Conference.

At the 1982 CIEC meeting, the ETD Executive Committee asked the Journal Study Committee to conduct a small survey of ETD members to solicit input on the need for a new journal and factors such as format, importance of indexing, and whether they would favor dues to help support the journal. Larry Wolf conducted the survey during the spring of 1982 and reported the results to the Journal Study Committee in a memo dated August 3, 1982.

Of the surveys returned, 208 favored ETD's publishing a new journal, 30 voted against, and 5 gave no response. Under the list of format suggestions, three items were rated the highest: contributed papers on technical subjects (but clearly engineering technology), contributed papers on engineering technology education, and editorial comment on issues affecting engineering technology. Concerning initiating annual ETD dues, 81 favored $5 per year, 41 favored $10 per year, 66 stated "whatever dues it would take to produce a quality journal," and 38 voted against dues. The same memo noted that Durward Huffman had been added to the Journal Study Committee as a two-year college representative.

On behalf of the committee, Larry Wolf submitted a formal proposal (dated October 1, 1982) to establish the *Journal of Engineering*

---

[1] Ken Merkel, "Editorial," *Journal of Engineering Technology* 1, no. 1 (March 1984): 3.

Technology to Gerald Rath, chairman of ETD, on November 4, 1982. A February 17, 1983, letter from Patricia Samaras, managing editor for ASEE, reported to Gerald Rath that the ASEE board had approved establishing such a journal. She went on to say,

> This decision reflects the confidence that the [b]oard, the Publications Committee and the staff have in the Engineering Technology Division. As one person put it, "The ETD is a viable and active [d]ivision — they're able to get things done."

At their June 1983 meeting, the ASEE board approved the $5 annual dues for ETD members, contingent upon documentation of a favorable ballot return from the members. In a letter dated July 20, 1983, the accounting firm of Deloitte, Haskins, and Sells of Houston, Texas, notified Larry Wolf that the ETD dues ballot passed: of 380 ballots received, 293 were in favor and 83 against. With the ASEE board's approval to establish the journal and institute ETD dues, the new editorial board for the *Journal of Engineering Technology* could begin work on the inaugural issue.

Now faced with the need to actually deliver the product, the committee was anxious as to what it might involve. Copyediting and printing production unsettled the engineering technologists on the Journal Study Committee no less than other engineers, with the inbred aversion that they have for such things. But, being dedicated people, infused with the vision of the *Journal of Engineering Technology* and possessing no small amount of courage, they were laid open to being conned into volunteering not only as an editorial board, but as a publishing house as well.

Wolf, who under his granny gown was the least gifted speller of all, had an ace in the hole: the she-Wolf, Barbara. Formerly the secretary to the editor of *Checkerboard Service* magazine of Ralston Purina and trained by the Sisters of St. Joseph, who were not only attired in, but sought to inculcate, "the old habits," Barbara was a walking dictionary and a perfectionist who would stalk any creative and artistic project to the kill.

Wolf organized the first team, with Ken Merkel agreeing to be the founding editor. Ken was a technical writer, the only one on the Journal Study Committee actually functioning within his comfort zone. Ken knew how to solicit papers and get a reviewing process started; he could be enticed to lead the editorial board if he thought the commitment was finite in time, since he was also working on his doctorate.

Wolf, a newly-appointed dean needing to open more opportunities for his faculty to survive and be promoted in a university aspiring to make it into the **top 20** among research universities, was sensitive to the fact that the *Journal* should not be viewed simply as a University of Houston project. For Wolf's purposes, the *Journal* would need to be independently refereed and ultimately indexed by the *Engineering Index*. Personnel and, ideally, the production office needed to rotate from one institution to another, important not only for scholarly credibility but also in order to get and maintain the financial support of ETD. The editorial board started as a floating crap game; the jobs would shift every two years. See Appendix F for a listing of *JET* editorial board members.

Keeping Barbara under wraps and smilingly diverting his colleagues' attention from the morass of production, Wolf volunteered to be the founding production editor. The other members of the board had to be spared, at least initially, the terror of production details. For in a rotation, they were each to be lured through the production editor position on their way to becoming editor-in-chief. Huffman, O'Hair, and Ron Scott signed on. But when no one was looking, Barbara was slipped in as the "editorial assistant" during the Wolf tenure. And Wolf would shrug and say, "See, there's nothing to it, guys."

Founding *JET* Editorial Board

Editor
Kenneth G. Merkel
University of Nebraska, Omaha

Associate Editor for Production
Lawrence J. Wolf
University of Houston, University Park

Editorial Board
Durward R. Huffman
Nashville State Technical Institute

Michael T. O'Hair
Purdue University, Kokomo

Ronald E. Scott
Wentworth Institute of Technology

Editorial Assistant
Barbara Wolf
Humble, Texas

In August of 1983, the *JET* board knew they had "arrived" when they received their new letterhead from Editor Ken Merkel.

The Inaugural Issue

The team in place, with dues having been approved to support the *Journal* but not a dollar anticipated for at least a year, Wolf started calling engineering technology schools for contributions to the printing and delivery of the first issue.

His first call was to Ted Kirkpatrick, a tireless fund-raiser, a former technical salesman, and president of Wentworth Institute. Wolf started with Ted on the theory that a true salesman is the easiest to sell. Ted committed Wentworth immediately for $1,000. That got the ball rolling. Thus encouraged, Wolf then shook down his own provost for another $1,000, in revenge for initiating "publish or perish" policies regardless of the consequences for technology faculty. By mid-morning, $5,000 was committed. Within a month, the editorial board had raised over $12,000. The inaugural issue came out promptly in March 1984, before any Engineering Technology Division member received a bill for the $5 dues assessment. That issue included the following note:

> *The inaugural issue of the* Journal of Engineering Technology *is made possible by initiation grants, without which this magazine would have been delayed until the Engineering Technology Division could generate sufficient*

revenue for publication. *The officers of the [d]ivision and the editorial staff of the* Journal *gratefully acknowledge the enthusiastic support of the following institutions and organizations:*

Arizona State University
    Department of Electronics and Computer Technology
Nashville State Technical Institute
Purdue University *School of Technology*
Society of Manufacturing Engineers
Southern Technical Institute
Technovate Corporation
Texas A & M University
    Department of Engineering Technology
University of Cincinnati *College of Applied Science*
University of Houston *College of Technology*
University of Houston *Office of the Provost*
University of Nebraska *Lincoln College of Engineering*
Wentworth Institute of Technology

They Always Judge a Book by Its Cover

Early in the life of the Journal Study Committee, Ken Merkel imposed upon Professor William Holmes of the University of Nebraska to hold a contest for the cover design in his "Introduction to Design" class. The committee had about 30 to choose from and selected two. Though professional looking, they were done by students and were not snappy enough for the *Journal*. The committee decided to look outside and actually pay some money.

Wolf contacted the St. Louis community college where he had been the associate dean and asked a commercial artist, Frank Stanton, for a price to do the job. Frank asked $250 for the cover and page designs. These would be delivered as mechanicals, which could be given directly to a printer. Frank chose the typefaces and conceived a way by which one four-color sheet could be folded to yield a color cover, rear cover advertising, and two interior color pages. He incorporated the ASEE logo into the second "O" of ENGINEERING TECHNOLOGY as a *Journal* logo. The *Journal* has not deviated from

Stanton's design to this day. And we have called upon his talents again for some of the design features of this publication.

This was an era when ASEE publications were looking tired. In fact, most engineering society journals of the day were very staid. When Wolf made presentations at ETD and ETC meetings, he was quick to pull out and pass around Stanton's cover mechanicals for the first issue, which featured a photo of the space shuttle. Most engineers had never seen a professionally designed mechanical. Neither had Wolf, but when his audience saw that cover, they envisioned quality and assumed that the committee knew exactly what it was doing when it came to publishing a magazine.

When the inaugural issue came out, everyone was delighted with its quality. A famous exchange took place between Michael Wald of the Fachhoshcule of Hamburg and Wolf at an ASEE meeting. Wald, editor of the *International Journal of Applied Engineering*, politely suggested that *JET*, with its aesthetic design and page spacing, was "Frankly, sort of empty." Wolf replied that he found the expensive *JAE*, published by Pergamon Press, to be "Frankly, sort of stuffy."

Production Development

The editorial board first tried a turnkey printer which would not only print but would do all the typesetting and mechanicals. But, since no journal of engineering technology had ever been produced before, estimates were wild with all sorts of expensive contingency clauses: "How much money have you got?" "Trust us." "Well, we know what we said, but this is what you get because this was extra and that was extra." Even though it was contracted as a turnkey job, Barbara had to communicate with the authors, proof the typesetting, and beg the printer to make corrections from the authors as well as those the printer originated, usually for extra money. Barbara sold the first ads and got the advertising mechanicals. She designed the mailing system and arranged the mailing.

So that ETD would not get financial cold feet, the second issue had to be supported from dues and advertising revenues. Instead of using a turnkey printer, Barbara copyedited each paper and took it to a typesetter, since typesetters do piecework by the page. She did the paste-ups right there on the family ping-pong table. When she got all of the ads together, she would put the paste-ups under her arm and seek a printer who was looking for a small job to squeeze in between larger production runs. When a printer could see exactly how much had to be done and that the issue was ready to roll, Barbara could tell if he really needed the work by his low bid. Successive issues came in within budget and well below the inaugural issue. Wolf concluded that printers were cost-effective when printing, not publishing.

Ten Years of Publication

The general format for the *Journal* has remained much the same for the first ten years of publication. One of the first pages in each issue contains the names and institutional affiliations of the editorial board, reviewers, and ETD officers. Since the *Journal* is primarily staffed by volunteers, credit to both individuals and institutions is important. The remainder of the issue is devoted to editorials, letters to the editor, an occasional invited paper, peer-reviewed papers, and fillers.

Indexing JET

Although the editorial board realized the importance of indexing for the *Journal*, it took approximately two years to receive approval. In a letter dated May 21, 1986, a representative from Engineering Information, Inc. notified Ken Merkel that, beginning with the March 1986 issue, the *Journal of Engineering Technology* would be abstracted and indexed in the *Engineering Index*.

Editorials

Originally, neither editorials nor letters to the editor were planned. But the board quickly changed its mind, since it had no good reason for the earlier policy. Besides, the board occasionally felt the need to lighten things up. Here is an excerpt from a fall 1986 editorial:

THIS DATA IS...!!!

And that's the problem. Data *may be Latin but it doesn't sound Latin because people don't pronounce it as Latin. If they did they would say "Dah-tah," not "Day-tah." If you don't use Latin pronunciation, why use Latin sentence structure and word endings?*

But, it was only after moving to Texas, which hasn't yet figured out whether it is the end of the South or the beginning of the West, that I really began to see the light. You see you can forget all that singular and plural if in fact the word is always singular. Really, have you ever heard anyone talk about a datum? Data *is a singular noun, never plural. If you want a* data point, *you always say just that.* Data *is a set of data points.*

In the South you may encounter many such singular nouns. To a northerner grits *may sound plural. But, who ever had just one grit for breakfast?* Grits *is a singular noun. And that's not a grit stuck between your teeth, it's a piece of grits, not unlike a data point.*

And, of course, you all *is singular. When you've been around long enough to learn the meaning from context you realize that* you all *means your entire self. When someone says, "You all have a good day!" that someone is wishing the day good for not only your physical body, but for your ego, your id, and on through your entire soul. When addressing more than one person from that context you would say naturally, "You all both," or "You alls." It's simple really.*

So combining what my ear says about Latin and what I've learned recently about nouns that are singular though they may look plural, I've edited this issue of the Journal. *Gleefully I changed Lyle McCurdy's "These data are" to "This data is." And, smugly I smiled as I left Ron Dilly's "This data is" gloriously untouched.*

But, upon writing this editorial, I am haunted by the feeling deep inside that someone, some printer's copy editor, or perhaps some typesetter will change it back and the one thing I've really worked for will be forever lost as Mike O'Hair takes over as Editor next year.

Larry Wolf
Editor

Letters to the Editor

Letters to the editor emerged quickly as individual readers took exception to published papers. The *Journal* let them butt and rebut to their hearts' content; that is until Charles Thomas of Purdue began publishing papers on the Myers-Briggs psychological indicators.

Suddenly the *Journal* struck a subject that every Ed.D. or Ed.D. candidate thought he or she was an expert on. The rest of us just watched the fireworks. Never mind that the debates had precious little to do with engineering technology. The letters to the editor and counter-letters grew to the size of the original papers. The board needed a policy about letters to the editor.

Much to the consternation of Thomas, who still had a few papers in the hopper, the board decided (in about five minutes) that *JET* would no longer publish papers on the Myers-Briggs indicators, a beautifully autocratic and subjective decision, considering that this was academia in the 1980's! But no more papers on that one subject meant no more run-on letters to the editor. And it worked. A collective sigh of relief went up.

Invited Papers

When important things happened, or a person of prominence was identified, or a particularly good piece appeared in the literature, the *Journal* would invite a paper. L. E. Grinter, of the famous Grinter Report, published the last paper of his long and distinguished career in the pages of *JET*'s inaugural issue.

We also scanned proceedings of the ASEE meetings. Authors of good papers were invited to spiff them up for the *Journal*. The board was criticized for this practice. Some wags called it re-publishing. The *Journal*'s rationale was that proceedings were, for the most part, unrefereed and definitely unindexed. The *Journal* staff saw a service to the author and the profession by giving the paper greater exposure and valuable jurying. In publishing, ET's needed every break they could get.

Peer-Reviewed Papers

Probably the most frequently asked question from potential authors is, "What kind of papers are you looking for?" The typical response includes such things as, "It must be of interest to our readers, engineering technology-related, timely, technically sound, and, of course, well-written." They then respond with, "OK. But what do you really want?" What most potential authors really want to know is what specific topics the board is looking for. The best indicator of topics or kinds of papers a journal will publish is the topics or kinds of papers they have already published. The *Journal of Engineering Technology* is no different. Potential authors are generally asked to examine the last three or four issues of the *Journal* to see which papers have survived the reviewing process.

But sometimes odd things happen. Michael O'Hair, as editor-in-chief, once received a paper from an ET faculty member that at first glance looked very strange: the author had submitted a manuscript with no capital letters. Thinking this was an oversight, O'Hair called the author. He was not available, so O'Hair explained the problem to his secretary. She quickly replied, "I'm glad someone else is concerned about his unwillingness to capitalize words." Apparently, this author did not believe in capital letters. O'Hair told the secretary to let Dr. Lower Case know that his manuscript would not be reviewed unless he capitalized appropriately. A revised manuscript arrived within the week, with no comments or explanations.

During the first nine years of publication (volume 1, number 1 through volume 9, number 2), 132 papers have been peer-reviewed. An inventory of papers by topic is listed below:

Topic	Number
Accreditation	2
Applied research	13
Computer usage	20
Curriculum	16
ET and industry	3
ET students	5
International ET education	5
Laboratory instruction	17

Topic	Number
Promotion/tenure/professional development	4
Studies of ET education	8
Teaching methods	15
Technical	17
Other	7
TOTAL	132

Conclusion

While the *Journal* appears in hindsight to have been well organized and even well conceived, it can, in fact, be better described as "subliminally organized." The secret of success is what other scholarly journals would pejoratively call a benign publication policy. But *JET* sees it as a market niche. *JET* accepted the fact that it was publishing for a readership which had few role models as contributors of papers. The *Journal* board provided a lot of encouragement, hand-holding, and a great deal of rewriting. Starch came into the process with a good panel of reviewers and with merciless rewriting on occasion, which the hapless author discovered only upon seeing the paper in print. Yet not a single author complained because the quality of the final result was obvious. And most authors realized that, without the existence and encouragement of the *Journal*, their work might not only have been denied the light of day and recognition of their peers, but in the majority of cases might not have been attempted at all.

Chapter 4

The James H. McGraw Award
by
Robert J. Wear (Retired)
Academy of Aeronautics

Since 1993 was the 45th year for the James H. McGraw Award as well as the 100th anniversary of the American Society for Engineering Education, it is an appropriate time to review and reflect on the people who have received the award and their accomplishments.

The James H. McGraw Award was a joint effort by ASEE and the McGraw-Hill Book Company. It all began in 1948, when Walter L. Hughes, a supervisor of admissions and placement at the Franklin Institute in Boston, wrote a letter to Edward E. Booher, a vice-president at McGraw-Hill. The letter suggested that the company should "give an award in the nature of a McGraw-Hill medal for an outstanding contribution in the field of Technical Institute Education."

 See Appendix I for photo captions and credits.

The ASEE annual meeting scheduled for the University of Texas that year, which would include a celebration of the 25th anniversary of the Technical Institute Division, would, Mr. Hughes thought, be an appropriate time for the award. He conceived the award as a one-time thing, just for that occasion. Mr. Booher replied, "It was a good idea and I am pursuing the matter of a McGraw-Hill award of some kind."

Later that month, James H. McGraw, Sr., died and company representatives decided that an annual award in his memory, for a member of the Technical Institute Division, was appropriate. The first award was not presented until the June 1950 meeting. Mr. Hughes, the award's proposer, did not live to see his suggestion come to fruition. A news release issued just before the meeting stated,

> *The award will consist of an annual prize of $500 and an appropriately engraved certificate. It is to be given for the purpose of recognizing and encouraging outstanding contributions to Technical Institute Education in the United States.*
>
> *Although the committee will consider contributions of various types, such contributions will lie within three categories—teaching, publications and administration. It is not expected that nominees for the award shall be outstanding in all three categories.*

The first chairman of the award committee was H. P. Rodes, assistant director, Relations with Schools, at the University of California at Los Angeles. The first award was presented at the annual banquet of the Technical Institute Division, at the University of Washington, June 1950.

Awardees[1]

1950 Seattle, Washington
Harry Parker Hammond, Pennsylvania State University

Harry Hammond was appointed as dean of the School of Engineering at Penn State and was responsible for starting the extension system and beginning the two-year college system. Prior to his administrative tenure at Penn State, he taught civil engineering at Penn State, Lehigh University, and the Polytechnic Institute of Brooklyn, where he served as professor and department head.

He made exceptional contributions to ASEE for over 25 years, chairing a number of important committees, including a special ECPD committee on technical institute-type education, and serving as vice-president and president. The main campus at Penn State has the engineering building named after him.

---

[1]Photographs and information regarding McGraw awardees and their institutional affiliations are current as of the year the award was conferred.

Harry Parker Hammond

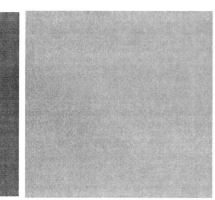

1951 East Lansing, Michigan
Robert Hoover Spahr, Pennsylvania State University

Robert Spahr was a Penn State faculty member who had served on the staff of the Society for the Promotion of Engineering Education, ASEE's predecessor. He was most noted, however, as the co-author of the 1931 report he prepared with William E. Wickenden, under the auspices of SPEE, called "A Study of Technical Institutes." This was a pioneer study of technical institute education and has been quoted many times in succeeding years.

He taught physics at the University of Kentucky, Kansas State Teachers College, and Wentworth Institute prior to chairing the Mechanical Engineering Department at the University of Maryland. At Penn State, he was in charge of field organization and supervised the engineering extension division.

The latter part of his career he was a member of the administrative staff at General Motors Institute, interspersing his industrial assignments with impressive and influential committee work in the areas of vocational and technical education.

Robert Hoover Spahr

1952 Hanover, New Hampshire
Arthur Lyman Williston, Wentworth Institute

Arthur Lyman Williston is considered the first to coin the term "technical institute," in 1922. He served on the faculty of MIT, Ohio State University, the Pratt Institute, and the Franklin Institute of Boston. He also consulted for Cooper Union. In 1910, Mr. Williston was called to Boston to supervise the development and organization of Wentworth Institute and served as Wentworth's first principal until his retirement. In addition, he consulted for federal government agencies, most notably the initiation and development of the Experimental Unit in Universal Military Training, Fort Knox, Kentucky, and served as a representative to education conferences in Edinburgh and Copenhagen.

He is the author of *Beyond the Horizon of Science*, which inspired the Arthur Williston Award for excellence in writing given by the Technical College Council, as well as numerous reports and articles on industrial and technical education.

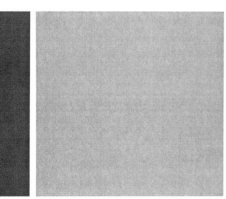

Arthur Lyman Williston

1953 Gainesville, Florida
Charles W. Beese, Pennsylvania State University

Charles W. Beese, another Penn State faculty member, served as the head of the Department of Industrial Education for four years. He then went into industry and returned to education as head of general engineering at Purdue. During World War II, he was director of war training at Purdue.

After the war, he directed Purdue's technical institute at Calumet and supervised the development of three more technical institutes in Fort Wayne, Indianapolis, and Michigan City. He later became director of Purdue's Technical Extension Division and eventually was named dean. He was instrumental in starting technical institute-type education in Indiana.

Due to his success at Purdue, Mr. Beese was asked to serve as education adviser for technical institute and industrial engineering programs in Japan and helped to plan a school for training executives in Turin, Italy.

Charles W. Beese

1954 Urbana, Illinois
Arthur C. Harper, Wyomissing Polytechnic Institute

Another Pennsylvanian, Arthur C. Harper, received his education at Penn State and went on to teach at Ohio State, the University of Illinois, and the Pratt Institute. In 1931, he became the director of Wyomissing Polytechnic Institute. Here he developed a program that expanded a trade school on a secondary level to the technical institute level. He was responsible for the organization of the entire technical institute curriculum, and he maintained close contact with graduates by initiating a follow-up system with cooperating industries.

He authored a number of articles and books in his field, in addition to preparing materials for studies and surveys significant to technical education, including material used in "A Study of Technical Institutes."

In 1942, he founded the American Apprenticeship Round Table while serving as chairman of the Technical Institute Division until 1946. Mr. Harper was active in the American Society of Mechanical Engineers and the Committee on Education and Training for the Industries.

Arthur C. Harper

1955 University Park, Pennsylvania
Frederick E. Dobbs, Ford Foundation

Frederick E. Dobbs came to Wentworth Institute from the Standard Oil Company of New York. At Wentworth, he served as a teacher, administrative assistant to the principal, and, during World War I, acting head of the school.

In 1924, Mr. Dobbs was named principal, and he served in that capacity until his retirement in 1952. His tenure is notable for his leadership in the development of the technical institute's facilities and his administrative skill in guiding Wentworth's curriculum development.

After his retirement, Mr. Dobbs served as a senior consultant for the Ford Foundation, where he was instrumental in building technical institute facilities and programs in Pakistan. On assignment in India, he developed training and research facilities for small businesses. He has been a guiding force in the development of technical institute education, both in the United States and abroad.

Frederick E. Dobbs

1956 Ames, Iowa
Charles S. Jones, Casey Jones School of Aeronautics

Charles S. (Casey) Jones was the first president of the Casey Jones School of Aeronautics, later the Academy of Aeronautics and now the College of Aeronautics. A former World War I flight instructor, he was involved in the technical training of aviation technicians through World War II and the Korean Conflict in governmental, private, and proprietary institutions.

He wrote numerous articles in his field, including a series of instruction articles for *Popular Science*, and published the *Casey Jones Cyclopedia of Aviation Terms*.

Mr. Jones served as TID chair from 1953-55 and was the division's representative to ASEE's General Council. He was a true pioneer in technical institute education for the aviation industry.

1957 Ithaca, New York
Arthur L. Townsend, Lowell Institute School

Arthur L. Townsend was a mechanical engineering faculty member at MIT from 1919 to 1944. He was associated with Lowell Institute School, a technical institute conducted under the auspices of MIT, and was named Lowell's director in 1944.

Charles S. Jones

Arthur L. Townsend

During World War II, he was appointed executive secretary of the planning committee for all Engineering, Science, and Management War Training activities at MIT and manager of the War Training Bureau. After the war, he served on a faculty committee to study revision of postwar courses at MIT, which led to the development of engineering technology courses at Lowell.

A member of a number of professional organizations, including ASEE, ASM, and SAE, Mr. Townsend chaired the mechanical and technical institute groups of ASEE's New England Section.

He wrote a number of articles on industrial safety and education. His entire professional career was marked by a concern for developing outstanding teachers who would instill in students both a sense of personal integrity and professional competence.

1958 Berkeley, California
Karl O. Werwath, Milwaukee School of Engineering

Karl Werwath was associated with the Milwaukee School of Engineering as a faculty member and administrator; in 1948, he was named president (his father preceded him in that position). Prior to teaching at MSOE, he received his degree in electrical engineering there, and furthered his education at Northwestern University and the University of Wisconsin.

Karl O. Werwath

He was very active professionally, serving in leadership positions in the National Council of Technical Schools. Mr. Werwath was well-known in ASEE as chairman of the Technical Institute Division, 1955-57, and as chairman of the English Speaking Union Technical Teachers Exchange.

A registered professional engineer in Wisconsin, Mr. Werwath served as a member of NSPE's Public Relations Committee and the Committee on Engineering Technicians, as well as being active in a number of business and civic organizations.

1959 Pittsburgh, Pennsylvania
Henry Preston Adams, Oklahoma State University

Henry P. Adams, director of the Technical Institute of Oklahoma State University, served on many ASEE committees as well as consulting for technical education in Pakistan. Many ASEE members in attendance at the 1979 meeting served under him in the latter programs. He was also involved with technical education programs in Indonesia.

At Oklahoma State, he was instrumental in the development of the technical institute, with seven curricula accredited by ECPD. His writings have appeared in a number of technical journals.

H. P. Adams is known for his outstanding contributions to technical institute education and for his talents as a teacher and administrator.

Henry Preston Adams

1960 West Lafayette, Indiana
Kenneth L. Holderman, Pennsylvania State University

Another Pennsylvanian, Kenneth L. Holderman, graduated from Penn State and worked for the Douglas Aircraft Company as a plant engineer before returning to Penn State. There he served in many capacities, including tenure as coordinator of Commonwealth campuses working with 14 technical institute-types throughout the state. He authored numerous articles published in the *Journal of Engineering Education* and *Technical Education News*.

I was chairman of TID when Ken was asked to present the award at an annual dinner. The dress for the affair was a white dinner jacket. I spoke to Ken about this, and he replied, "I don't have one with me and I don't intend to rent one." I told him that all the other platform persons would be wearing one, so it was up to him. The night of the dinner, I was a bit apprehensive about his appearance but was relieved to see him come in wearing a white jacket. When I spoke to him, he said that he had borrowed it from Eric Walker, then president of Penn State.

Ken served on many ASEE committees and was chairman of the Technical Institute Division.

Kenneth L. Holderman

1961 Lexington, Kentucky
H. Russell Beatty, Wentworth Institute

H. Russell Beatty graduated from the University of Maine, went into industry, and returned to education at the Pratt Institute as a faculty member and then acting dean; he was responsible for both four-year engineering and technical institute programs. He left Pratt for Wentworth Institute of Technology and became its president in 1953, earning a reputation for initiative and imagination.

A tireless worker in ASEE, he chaired three divisions — Industrial Engineering, Educational Methods, and Evening Engineering Education — and was a member of numerous other committees.

1962 Colorado Springs, Colorado
Eugene H. Rietzke, Capitol Radio Engineering Institute

Eugene Rietzke was founder, president, and chairman of the board at Capitol Radio Engineering Institute. He was co-founder of the National Council of Technical Schools, which later integrated into ASEE. He was instrumental in the Technical Institute Sub-committee of ECPD, which laid the groundwork for accreditation of the technical institute curriculum.

H. Russell Beatty

Eugene H. Rietzke

As a spokesman for technical institute education, he represented many educational organizations in testimony before Congressional committees and consulted to governmental committees involved in forming legislation which would affect technical institute education.

1963 Philadelphia, Pennsylvania
Lawrence V. Johnson, Southern Technical Institute

Larry Johnson came from the Georgia Institute of Technology, where he taught physics and aeronautics, and served as the acting director of the Daniel Guggenheim School of Aeronautics. He spent one year teaching at the American Biarritz University in France, and then became the director of the Georgia Tech technical institute program. This resulted in the establishment of Southern Technical Institute in 1948. For a real tour of the campus in the late 70's, Larry was an excellent guide, since he knew it all and probably planned it all, from classrooms to broom closets.

He served on a number of TID committees and was active in accreditation through ECPD.

Lawrence V. Johnson

1964 Orono, Maine
A. Ray Sims, University of Houston

A. Ray Sims taught and was in administration in Texas for 13 years. He then joined the University of Houston to direct the in-plant training of the Reed Roller Bit Company. In 1945, he became part of the Training within Industry Service of the War Manpower Commission and the next year became dean of the College of Technology, University of Houston.

He chaired the Technical Institute Division from 1959-61, was active in ECPD, and was a member of a delegation of technical institute administrators who went to the Soviet Union in 1962. He also served as a consultant for polytechnical education in industry.

1965 Chicago, Illinois
C. L. Foster, Central Technical Institute

C. L. Foster, president of Central Technical Institute in Kansas City, began his career as an instructor at the First National Television School, which later became Central Technical Institute. In 1952, he became president.

One of the pioneers of television, he was associated with the construction and operation of an experimental television station licensed

A. Ray Sims

C.L. Foster

in 1934. He became the chief instructor in 1937 and developed curriculum to train commercial airline radio operators. He was active with IEEE, as well as ECPD and the National Council of Technical Schools. In 1963, he was awarded honorary membership in Tau Alpha Pi.

1966   Pullman, Washington
Walter M. Hartung, Academy of Aeronautics

Walter M. Hartung, president of the Academy of Aeronautics, graduated from New York University and began his career as designer of light aircraft and went on to design racing planes before joining the Casey Jones School of Aeronautics. There he was active in the development of engineering curriculum and the design of aircraft. During World War II, he was in the Air Transport Command in the European and Pacific theaters.

After the war, he returned to the Academy of Aeronautics, a successor to the Casey Jones School. He became executive vice-president, dean, and eventually president. He was active in many organizations, such as AIAA, and served as mayor of his town. Mr. Hartung was chairman of the Technical Education Delegation that went to the Soviet Union in 1961.

Walter M. Hartung

1967 East Lansing, Michigan
Cecil C. Tyrrell, New York Institute of Applied Arts and Science

Cecil C. Tyrrell began his career teaching at the University of Maine, then moved to Pratt Institute and Rutgers University. In 1946, he was appointed chief administrative officer at the New York State Institute of Applied Arts and Science at Binghamton, later renamed Broome Technical Community College.

At Broome Tech, he worked on developing two-year programs with courses designed to upgrade skills of the disadvantaged people of Binghamton, cooperative programs with industry, and placement services for graduates.

He was a long-time member of ASEE and chairman of the steering committee that resulted in the 1962 ASEE report, "Characteristics of Excellence in Engineering Technology Education," sometimes called the McGraw Report. In addition to his many committee assignments in ASEE, he was active in community affairs.

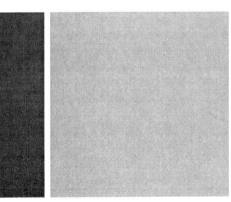

Cecil C. Tyrrell

1968 Los Angeles, California
William N. Fenninger, New York State Department of Education

William Fenninger, who was executive secretary of the American Technical Education Association when he received the McGraw Award, earned degrees at Ohio State University and at Franklin and Marshall. He taught at Ohio State and the Pratt Institute before going into industry, serving as educational director for the Brooklyn Edison Company.

He returned to the academic scene as head of the Electrical Department at Rochester Athenaeum and Mechanics Institute, now known as Rochester Institute of Technology. From there he joined the New York State Department of Education.

During World War II, he supervised national defense courses taught in New York's engineering colleges and state technical schools. After the war, he planned the electricity and chemistry courses for engineering technicians at New York's agricultural and technical institutes and was later promoted to chief of the Bureau of Trade and Technical Education.

After his retirement, he joined the American Technical Education Association as their first executive secretary. In addition to serving as chair and planner of AETA's annual regional conferences, he also wrote 63 issues of its official paper, *Technical Education Newsletter*.

William N. Fenninger

1969 University Park, Pennsylvania
Winston D. Purvine, Oregon Technical Institute

Winston Purvine taught in Oregon before becoming the director of a special school for the rehabilitation of World War II veterans. This school became Oregon Technical Institute, and Winston was named president in 1961. Oregon Tech became a four-year institution under his direction.

Winston was active in ASEE and ECPD, serving as chairman of the division and as a member of the Technical Institute Administrative Council. In addition to academic activities, he has been involved and held many positions with charitable and community service organizations.

1970 Columbus, Ohio
Hugh E. McCallick, University of Houston

Hugh McCallick received the McGraw Award for his leadership in the origin, development, and establishment of four-year baccalaureate degree programs in engineering technology; for his contribution to the improvement of technology education in private, private proprietary, and public educational institutions; for his devotion to the standards of quality implicit in the accreditation requirements of the Engineers' Council for Professional Development; for his exceptional record of accomplishments as dean of the College of Technology, University of

Winston D. Purvine

Hugh E. McCallick

Houston; and as a consultant to business, industry, government, and educational institutions at home and abroad. He has been active in many areas of ASEE.

Dr. McCallick attended Texas College of Mines and Metallurgy, Texas Western College, the University of Houston, and the University of Texas. His degrees are in technical industrial management and institutional administration.

He served in the Navy during World War II and then joined the faculty of the University of Houston. He was the executive vice-president of Holmes Institute and a member of the board of the Institute of Computing Sciences.

1971 Annapolis, Maryland
Melvin R. Lohmann, Oklahoma State University

Melvin R. Lohmann was presented the McGraw Award for his efforts to improve the quality and extend the level of engineering technical education, for his guidance in the founding of Oklahoma City Technical Institute and the inauguration of the baccalaureate degree in engineering technology at Oklahoma State University's School of Technology, for his enterprise in furthering engineering and technical education in many countries around the world, and for his leadership in ASEE and ECPD. He was professor of industrial engineering and

Melvin R. Lohmann

dean at the College of Engineering, Oklahoma State University. He holds degrees from the University of Minnesota, University of Pittsburgh, and University of Iowa.

He began his career as an industrial engineer at the Aluminum Company of America and a part-time instructor in the evening extension of Penn State. In 1941, he was appointed assistant professor of industrial engineering at Oklahoma State University. His tenure there was interrupted by World War II, when he served in the South Pacific in the Marine Corps.

After the war, he returned to academia at Oklahoma State University, where he worked to develop its technical institute.

He served as a consultant to programs in Pakistan, as well as to various assistance programs in Latin America, Africa, and Europe. A past president of ASEE and ECPD, he contributed many articles to professional journals and was an active consultant to industry.

1972 Lubbock, Texas
Richard J. Ungrodt, Milwaukee School of Engineering

Richard J. Ungrodt was cited for his leadership in the development of engineering technology programs at both the associate and baccalaureate levels; for his years of teaching engineering students, technologists, and technicians; for his administration as vice-president

Richard J. Ungrodt

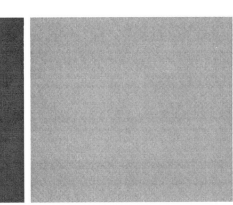

for academic affairs at the Milwaukee School of Engineering; and for his overall commitment to excellence in service to ECPD and ASEE.

He graduated from the Milwaukee School of Engineering and was employed by Allis-Chalmers Manufacturing Company in Milwaukee while continuing graduate study at Illinois Institute of Technology.

In 1947, he returned to MSOE, serving as instructor, assistant and associate dean of engineering, dean of engineering, and finally vice-president for academic affairs.

In addition to his teaching and administrative duties at MSOE, he has been active in many professional, religious, and civic organizations and involved in many positions with ASEE. He was director of the ECPD board.

He was been involved regularly as a speaker for civic groups, educational associations, engineering and technical societies, and other such organizations.

1973   Ames, Iowa
G. Ross Henninger, Ohio College of Applied Science

G. Ross Henninger received the McGraw Award for his leadership in engineering technology. He was a member of the Committee of 21 and served as the director of the ASEE national survey of technical institute education that produced the book *The Technical Institute in America*.

G. Ross Henninger

He was president of Ohio College of Applied Science and its predecessor, Ohio Mechanics Institute.

1974 Troy, New York
Merritt Alvin Williamson, Vanderbilt University

Merritt A. Williamson, distinguished professor of engineering management at Vanderbilt University, graduated from Yale and holds master's degrees from Yale, California Institute of Technology, and the University of Chicago, as well as a doctorate from Yale.

He served in the Navy and held several industrial positions before becoming dean and professor of engineering at Penn State. He was best known for his leadership in educational administration. At Penn State, he organized the engineering technology unit. Dr. Williamson chaired a joint ASEE/ECPD committee to clearly define engineering technology and developed a program for the certification of engineering technicians and the Institute for the Certification of Engineering Technicians.

He was active in ASEE as chairman of various organizations, including the Administration of Research and the Engineering Division of the Land Grant Association; he was president of the Yale Engineering Association and ASEE.

Merritt Alvin Williamson

1975 Fort Collins, Colorado
Louis J. Dunham, Jr., Franklin Institute of Boston

Louis J. Dunham graduated from Harvard University and entered the Army, where he received diplomas from the University of Virginia and Yale University. He attended the Franklin Institute and began his career there as an instructor. He developed texts in electrical engineering, became a registered professional engineer, and was a consultant and visiting professor in Venezuela.

In 1957, he was elected director of the Franklin Institute. He was responsible for the expansion of the physical plant and upgrading the level of education offered.

He had many professional memberships and held offices in ECPD. He was chairman of the McGraw Award Committee of ASEE, the National Association of Foreign Student Advisors, and the Massachusetts Society of Professional Engineers; the vice-president of the National Council of Technical Schools; and secretary and vice-chairman of ASEE's Technical College Council.

Louis J. Dunham, Jr.

1976 Knoxville, Tennessee
Eugene Wood Smith, Cogswell College

Eugene Wood Smith received his baccalaureate degree from California Institute of Technology and went to work as a draftsman for the Los Angeles County Surveyor's Office. In 1930, he moved to the academic area, becoming an instructor at Cogswell College, where he developed the curriculum in mechanical engineering technology. The first vice-president and the president of the college, he expanded the faculty and obtained necessary equipment for laboratories, resulting in ECPD's 1951 accreditation for electronic technology.

He was involved nationally in the development of engineering technology education as a member of the Committee of 21, 1953-59. He was an officer of the Technical Institute Division as well as the Technical Institute Council, and he was a member of the steering committee which developed criteria and data for the National Science study included in the Grinter Report.

He was active in California, serving on ASEE committees and as a consultant for many California colleges. He was well-known as a fine storyteller, with stories for any occasion.

Eugene Wood Smith

James H. McGraw Award

1977 Fargo, North Dakota
Donald C. Metz, University of Dayton

Donald C. Metz had a career that spanned the fields of engineering and education. He began at Purdue University with a bachelor of science in electrical engineering and a master of science in industrial engineering. His industrial experience started as a student engineering trainee for General Motors, and then he went to Hughes Heating and Air Conditioning Company. In 1940-46, he served in the Army. After the war, he returned to Purdue as supervisor of electrical technology and then as head of the technical institute. In 1957, he was selected by the University of Dayton to develop their technical institute.

He worked for accreditation. This was accomplished for the electrical, industrial, and mechanical curricula in 1954.

He served in many areas of ASEE, was a registered professional engineer in Ohio, and was the first American principal in Nigeria. He completed his career as associate dean of engineering technology at Southwest Minnesota State College, where he was known as "Mr. Engineering Technology."

Donald C. Metz

1978 Vancouver, British Columbia, Canada
Joseph J. Gershon, Bell and Howell School

Joseph J. Gershon had a career devoted to technical institute education. While working towards a bachelor of science in electrical engineering, he was an instructor in radio communications at the Bell and Howell School in Chicago. From 1946 to 1953, he headed the Department of Advanced Technology. In 1954, he became the director of the Resident School, then dean and vice-president. In 1971, he was appointed senior vice-president of the Bell and Howell School system.

He was able to put the proprietary school on equal footing with the academic institutes in the accreditation process with ECPD and successfully promoted mathematics and humanities in the curriculum of technical institutes.

Throughout his career, he was involved with ASEE and ECPD, as well as with numerous local, state, and national committees concerned with technical institute education.

1979 Baton Rouge, Louisiana
Robert J. Wear, Academy of Aeronautics

Robert J. Wear was honored with the 30th McGraw Award for his dedication to the advancement of engineering and technology educa-

Joseph J. Gershon

Robert J. Wear

tion as a teacher and administrator, as a member and officer of national societies and councils, and for his contribution to panels, technical teacher exchange, and national committees.

He received his bachelor of science in industrial engineering from Fairleigh Dickinson University and a master of arts from New York University. He was involved with teaching in aeronautical engineering technological areas at the Casey Jones School of Aeronautics, later the Academy of Aeronautics. He became director of the Day Division, academic dean, and administrative dean.

1980 Amherst, Massachusetts
Lyman L. Francis, University of Missouri

Lyman L. Francis has been closely associated with engineering technology since the late 1940's through his activities in manufacturing. In 1965, he was elected to membership in the Engineering Technology Committee of the Engineers' Council for Professional Development. He demonstrated outstanding leadership by visiting more than 40 campuses as a curriculum evaluator, evaluation committee chairman, and consultant.

Professor Francis was one of the first Society of Manufacturing Engineers' representatives to the ECPD Board of Directors. He served as chairman of the ECPD Council, the first representative from engineering technology to hold such a position.

Lyman L. Francis

He was a member of various committees in the Engineering Technology Division of ASEE and presented numerous papers at regional and national meetings of the society.

1981 Los Angeles, California
Ernest R. Weidhaas, Pennsylvania State University

Ernest R. Weidhaas was assistant dean for Commonwealth campuses at Penn State University. After serving in the Army in World War II, he graduated from New York University with bachelor's and master's degrees in mechanical engineering. He joined the faculty at the University of Maine in 1953, which led to the publication of the McGraw-Hill series, "Creative Problems in Engineering Graphics."

In 1954, he was asked to join the faculty at Penn State University, where he administered the Engineering Graphics Division. In 1965, he was appointed head of the General Engineering Department as well as the engineering technology units at the 18 Penn State campuses. He was the director of 12 associate degree engineering technology majors at the 18 Commonwealth campuses.

He made great efforts to increase female and minority applicants in engineering and engineering technology. He has served as the chairman of ASEE's Technical College Council.

Ernest R. Weidhaas

1982 College Station, Texas
Michael C. Mazzola, Franklin Institute of Boston

Michael C. Mazzola was involved in engineering technology since the Committee of 21, a predecessor of ASEE's Engineering Technology Council. Michael began his formal post-secondary education in 1946 at the school where he later became president, the Franklin Institute. He graduated in 1948 with a certificate in structural design and architecture. He received a master's degree from Harvard in 1951 and is a registered professional engineer in the Commonwealth of Massachusetts.

He began employment as a civil engineer and started to teach at Franklin, progressing from faculty, to associate dean of faculty, to dean of faculty. He was appointed director in 1975, and in 1981 he became the first president of the institute.

His many professional activities included ASEE, the American Society of Civil Engineers, and the New England Association of Schools and Colleges, in addition to membership in the Committee of 21. He served as vice-chairman and chairman of the Technical College Council and was on the ASEE Board of Directors.

Michael C. Mazzola

1983 Rochester, New York
Walter E. Thomas, Florida International University

Walter E. Thomas received his undergraduate and master's degrees from Purdue University, and he taught at several engineering schools before entering industry. In 1964, he was appointed professor and head of the Manufacturing Engineering Technology Department at Purdue University. He developed associate degree programs at seven Purdue and Indiana University campuses and two-year, add-on programs at four Purdue campuses.

In 1973, he became director of the Division of Engineering Technology and the Division of Industrial Technology at Florida International University, and the following year he was appointed associate dean of the School of Technology. In 1974, he received his Ph.D. in education from Indiana Northern University.

He served as chairman of the Engineering Technology Division and on many committees for the International Society of Manufacturing Engineers. He has qualified as a certified manufacturing engineer for life.

Walter E. Thomas

1984   Salt Lake City, Utah
Stephen R. Cheshier, Southern College of Technology

Stephen R. Cheshier was recognized for excellence as a teacher, author, and administrator. His excellence as a teacher was demonstrated by his rapid rise in the academic ranks, progressing, in only five years, from instructor to full professor and department head at Purdue University. He has won six awards for outstanding teaching at the military, vo-tech, college, and university levels.

From 1971 to 1980, he was on the faculty at Purdue and served on many major committees. From 1976 to 1980, he was department head and had administrative responsibility for programs at two Purdue regional campuses.

In 1973, he created the Evening Adult Program and electronics curriculum for Indiana's Vocational College and was the program administrator for three years.

He was selected as president of Southern College of Technology in Marietta, Georgia. He was a two-time chairman of the Engineering Technology Leadership Institute and head of the Georgia University System Senior College President's Organization. Southern Tech has grown under his leadership, acquiring funding for two new buildings, several major grants, and four new degree programs, as of the award year.

Stephen R. Cheshier

1985 Atlanta, Georgia
James P. Todd, Westland College

James P. Todd received the McGraw Award in recognition of his 25 years of service to engineering technology education.

Since 1984, he has been the chief administrative officer at Westland College, a private two-year college in Fresno, California. Prior to holding this position, he was president of Vermont Technical College; he was instrumental in initiating an associate degree program for IBM's Burlington plant, where the day courses were taught by college faculty. External degree programs in electronic and mechanical engineering technology were offered in the evening for employees.

Mr. Todd has taught at California State Polytechnic University at Pomona and Bridgeport Engineering Institute. His industrial experience included positions at Giannini Scientific Corporation, AVCO Lycoming Division, Aerojet General Corporation, Southern Pacific Company, Pratt & Whitney Aircraft, and Cal Tech's Jet Propulsion Laboratory.

He has been a long-time member of ASME, AIAA, ASEE, and ASTM. In June 1986, he became chairman of ABET's Technology Accreditation Commission.

James P. Todd

1986 Cincinnati, Ohio
Anthony L. Tilmans, Kansas Technical Institute

Anthony L. Tilmans received the McGraw Award for his excellence as a teacher and administrator and for his leadership in engineering technology education.

He held several positions at the main and Johnstown campuses of the University of Pittsburgh, Indiana State University at Evansville, and California State Polytechnic University at Pomona. At Indiana State, he developed the curricula for associate and baccalaureate programs. He served as the provost at Wentworth Institute of Technology and then became president of Kansas Technical Institute at Salina.

He has been an active member of ASEE and chaired the Engineering Technology Division when the *Journal of Engineering Technology* began publication.

Co-founder of the Engineering Technology Leadership Institute, he has served on the executive council of ETLI since its founding. He has been an *ad hoc* visitor of ABET's Technology Accreditation Commission and chairman of a visiting team for 12 institutions.

He is a registered professional engineer and active in professional organizations. As a member of the American Society of Civil Engineers since 1959, he has worked at local and state levels.

Anthony L. Tilmans

1987 Reno, Nevada
Lawrence J. Wolf, University of Houston

Lawrence J. Wolf began his teaching career in 1964 at St. Louis Community College at Florissant Valley, Missouri, where he was the founding chairman of the Mechanical Engineering Department. He went on to teach at Wentworth Institute of Technology and Purdue University, Calumet. In 1980, he was appointed dean at the College of Technology, University of Houston, where he was responsible for upgrading the curriculum and introducing graduate education. His position as an educator has taken him to such locales as Saudi Arabia, Norway, and Singapore.

He is a long-time member of ASEE, serving on numerous committees and holding a variety of offices. He chaired the founding committee for the *Journal of Engineering Technology*.

Dr. Wolf has also been an active member of the Technology Accreditation Commission of ABET. In 1986, he was elected charter president of the Texas Association of Schools of Engineering Technology. He has also been appointed to commissions for the American Society of Mechanical Engineers, the National Academy of Engineering, and the coordinating board of the Texas college and university system. He established and served as faculty advisor for Tau Alpha Pi and is the author of numerous books and articles. [In 1991, Dr. Wolf was named president of Oregon Institute of Technology.]

Lawrence J. Wolf

1988 Portland, Oregon
Harris T. Travis, Southern College of Technology

Harris T. Travis began his teaching and counseling career at Purdue University, West Lafayette campus, after 12 years as an engineer with the Navy. He completed his 10-year affiliation with Purdue as professor and chairman of the Mechanical Engineering Technology Department and moved to Southern College of Technology in Marietta, Georgia.

At Southern College of Technology, he established a college-wide retention program, three new undergraduate programs, and two graduate programs. He played a significant role in the transition from a branch campus of Georgia Institute of Technology to a senior college within the Georgia university system.

Dr. Travis has served on ASEE's Board of Directors, as ETD chair and vice-chair, and as chairman of ETCC's Committee on Minorities and Females in Engineering Technology, in addition to tenure as program and session chair for a number of ASEE conferences. He has consulted in England, Singapore, and Brazil.

Harris T. Travis

1989 Lincoln, Nebraska
Edward T. Kirkpatrick, Wentworth Institute of Technology

Edward T. Kirkpatrick, McGraw's 40th award recipient, began his career in engineering technology as an assistant professor at the Carnegie Institute of Technology at Pittsburgh. He then went to the University of Pittsburgh, where he became department head of mechanical engineering. He was the founding director of the computing center at the University of Toledo, and at Rochester Institute of Technology he served as dean of the College of Technology.

Dr. Kirkpatrick was chairman and primary author of the engineering technology education section of the American Academy of Engineering's study, "Engineering Education and Practice in the United States."

At Wentworth, he was responsible for the rise in enrollment by encouraging the faculty to write textbooks and develop four-year engineering technology programs, as well as a weekend college, cooperative education, and several international projects.

He has participated in many organizations: a member of the Engineering Manpower Commission, chairman of ASEE's Engineering Technology Council, a member of ASEE's Board of Directors, and the ASEE representative on ABET.

A former practicing engineer, Dr. Kirkpatrick was named an ASEE Fellow in 1983.

Edward T. Kirkpatrick

1990 Toronto, Ontario, Canada
Ray L. Sisson, University of Southern Colorado

Ray L. Sisson, dean of the College of Applied Science and Technology at the University of Southern Colorado in Pueblo, began teaching electronics in 1953 while in the Navy. In 1960, he received his bachelor's degree in electrical engineering from the University of Colorado, Boulder, and was named head of the Engineering Department. He went on to earn his master's degree in electrical engineering from Colorado State University, Fort Collins, and his doctorate from the University of Northern Colorado.

At the University of Southern Colorado, he has been responsible for introducing more than 20 new courses and for the transition from a vocational electronics program to an electronics engineering program. After his appointment as dean, he led and supervised a similar transition from vocational to engineering technology programming in civil and mechanical engineering technology, as well as the creation of two new two-plus-two degree programs in manufacturing and metallurgical engineering technology.

In addition to his achievements locally, he has represented engineering technology at the state and national levels. He had the University of Southern Colorado designated as Colorado's only polytechnic university. He was also active in placing USC under the Land Grant College Board and the State Board of Agriculture, creating the

Ray L. Sisson

Colorado Advanced Technology Institute, and acquiring funds for constructing a new technology building on the USC campus.

Dr. Sisson has been active in ASEE affairs, serving on many committees as well as holding all the offices within ETC.

And, incidentally, he was the first recipient of the McGraw Award to raise his voice in song at the annual banquet.

1991 New Orleans, Louisiana
Earl E. Gottsman, Capitol College

Earl E. Gottsman has spent virtually all of his career with engineering technology students. During his full-time teaching, he developed and taught courses in physics and mathematics. He has led Capitol College as professor, dean, and vice-president.

In 1990, he initiated a master of science program in systems management. As vice-president, he has more than tripled undergraduate offerings, adding curricula at the two- and four-year levels, including two- and four-year programs in telecommunications. And, for each curriculum, he created a separate and independent industrial advisory board.

He has served ASEE in many ways, such as being the first chairman of the Frederick J. Berger Award Committee, ETD chair, a member of the ETC Board of Directors, and through his membership on many committees.

Earl E. Gottsman

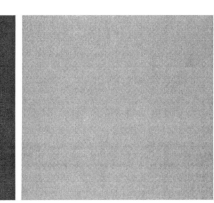

1992   Toledo, Ohio
Frederick J. Berger,   Tau Alpha Pi Executive Director

Frederick J. Berger was the founding executive director of Tau Alpha Pi, a national honor society for technology students. In 1960, he began teaching engineering and engineering technology at City University of New York, Bronx Community College.

With an industrial background, as early as 1962 he developed a programmed instructional approach to electronics and was instrumental in the preparation of films on basic electricity and electronics. He obtained grants that enabled him to implement computerized instruction in the mechanical area and to incorporate electronics and semi-conductor device laboratory techniques in the electrical field. He set up IBM personal computers for use by engineering technology and business departments and instructed the students in repair. He established a liaison with four-year institutions for a smoother transfer of technology students to the baccalaureate program and was a member of the chancellor's committee that planned a B.S. program in engineering technology at City College of the City of New York.

His efforts extended to many professional activities: Tau Beta Pi, IEEE, American Society of Nuclear Engineers, and ECPD/ABET, serving on many evaluation teams. As a member of ASEE, he served on the Nominating and Manpower Committees, Relations with Industry

Frederick J. Berger

Division, Engineering Technology Leadership Institute, and the Engineering Technology Division. In 1990, ASEE inducted him as a Fellow.

He has been deeply involved with Tau Alpha Pi, which has grown into over 135 chapters nationally and has become an essential part of engineering technology. He established the *Tau Alpha Pi Journal*, which he edits and publishes annually.

1993 Champaign, Illinois
George A. Timblin, Central Piedmont Community College

George A. Timblin, current head of the Engineering and Advanced Technology Department at Central Piedmont Community College in Charlotte, North Carolina, began his career in engineering technology with a bachelor's degree in electrical engineering from Duke University in 1962. Prior to his teaching career, he worked for a number of years in industry, including experience at Duquesne Light Company, Boeing, and the Lockheed-Georgia Company. He accepted a position as instructor of electrical and electronics technology at Central Piedmont in 1970 and received his master's degree from the University of North Carolina, Charlotte, in 1975.

Mr. Timblin has served Central Piedmont in a variety of leadership roles, such as director of electrical and electronics engineering

George A Timblin

technology programs and head of the Technology Department. All six engineering technology programs have received accreditation during his tenure.

As head of the Engineering and Advanced Technology Department for 14 years, Mr. Timblin has contributed to the doubling of programs, enrollment, and full-time faculty. He continues to be instrumental in establishing and maintaining beneficial partnerships with industry. Under his guidance, Central Piedmont has become one of the first 48 institutional members of the IBM CIM in Higher Education Alliance.

His professional affiliations include IEEE, the North Carolina Society of Engineers, and the Charlotte Engineers Club. In ASEE, he has served as chair of the Engineering Technology Division and has chaired the Berger Committee. He is currently Central Piedmont's representative to the Engineering Council and is board member/chairman of the Engineering and Technology Nomination Division.

Chapter 5

ETD/ETC Minutes

Compiled by
Frank A. Gourley, Jr.
West Virginia Institute of Technology

Following are quotes excerpted from minutes and reports of engineering technology meetings within ASEE over the years. They are arranged in chronological order, with paragraphs indicating separate topics within each meeting. Each meeting is identified separately. Missing meetings are generally an indication that minutes were not available. Items reported in both Executive Committee and business meetings are usually quoted only once. See Appendix G for tables of ETD minutes topics. The compiler's comments are indicated by brackets. Minutes have been edited for consistency in spelling, punctuation, capitalization, and numerals.

 See Appendix I for photo captions and credits.

Part I: Excerpts from Division Meetings
1949-1992

June 1949 Troy, New York
TID Business Meeting

Walter L. Hughes, retiring chairman of the Technical Institute Division, and Karl O. Werwath, president of the Milwaukee School of Engineering, reported on the formation and recent activities of the Technical Institute Division's National Committee of 21, including the committee's interim meeting at Milwaukee in December 1948.

A report on one of the activities mentioned above was presented by Henry P. Adams, chair of the Sub-committee on Divisional Reorganization. After considerable discussion at two meetings of the Technical Institute Division, a revised series of by-laws was finally adopted unanimously.

Dean H. P. Hammond of Pennsylvania State College reported that more than 25 curricula of technical institute-type institutions have been accredited by the Engineers' Council for Professional Development, to date.

Harold T. Rodes of the University of California reported for the James H. McGraw, Sr., Award Committee. The James H. McGraw, Sr., Award for distinguished contribution to technical institute education will be made for the first time in June 1950.

At the joint session with the Junior College Division, it became clearer than ever before that, despite some differences in organization and educational philosophy, the Technical Institute and Junior College Divisions of the society have a great deal in common. The Technical Institute Division voted to continue the National Committee of 21, in the hope that this nucleus would provide the leadership and continuity essential to the technical institute movement.

June 1950 University of Washington Seattle
TID Business Meeting

It was agreed that those institutions wishing to do so would forward their technical institute exhibits on display at the Seattle meeting to Purdue University for the Technical Institute Workshop to be held there from July 17 to 28, 1950.

E. E. Booher was elected to fill out the term of the late W. L. Hughes. [The minutes of this meeting included a four-page listing of the 93 people attending the division meetings, along with their titles, their affiliations, and addresses. Also included were the results of a survey of reorganization of the Technical Institute Division, conducted by H. P. Adams. The purpose of the survey was to make recommendations concerning the tentative reorganization plans of the division originally submitted by Mr. Hughes, division chair. The conclusions of the survey were that the Technical Institute Division (1) continue representing members of institutions offering technical institute-type training, auxiliary to, but not in the field of, professional engineering; (2) continue to operate as a division of ASEE and ECPD; (3) continue to be called the Technical Institute Division; and (4) continue to promote and develop engineering technology education.]

June 1952   Dartmouth College   Hanover
TID Business Meeting

Jeanne Miller with *Technical Education News*, McGraw-Hill Book Company, reported that the Technical Institute Studies Committee has compiled a bibliography of about 350 titles in three categories — articles, books and pamphlets, and graduate studies. Each category is further classified according to subject, for instance, history of technical institutes, organization, administration, curriculum building, guidance, placement, etc.

C. W. Beese, Purdue University, chairman of the Relations with Industry Committee, reported that the technical institute needs to sell its programs to prospective students and acquaint other important groups with this type of education. Some of the more important ways that this can be done are through joint committees of technical institute administrators, teachers, and representatives of industry, through which special curricula can be developed and cooperative plans, inspection trips, industrial opportunities, industrial teachers, industrial information relationships with engineering and other societies and follow-ups on graduates can be initiated. [This set of minutes listed members of the National Committee of 21 and committee chairs.]

June 1953 University of Florida Gainesville
TID Business Meeting

R. W. Marsh, SUNY-Buffalo, reported on activities of the Completion Credentials Committee that included a resolution as follows: "Be it resolved that, in interest of clarity, public understanding of the nature of technical institute training, and the place of the technician in industry, all members of the ASEE be urged to make reference to graduates of technical institutes as engineering technicians and advocate a designated associate degree, as the appropriate title of accomplishment."

C. S. Jones, Academy of Aeronautics, reported that the Cooperation with Office of National Defense Committee had continued its contacts with government agencies urging the establishment of a reserve technical training corps in technical institutes.

Cecil C. Tyrrell, New York State Institute of Applied Arts and Science, Binghamton, New York, reported for the Student Selection and Guidance Committee. A summary of the study made by this committee indicates a relatively large number of schools, 15 out of 39, were doing no testing of their own for the determination of the mental ability of their students at admission. Studies of the information that was gathered indicate that the high school record still is the most reliable indicator of success in our programs, particularly when analyzed in relation to the high school courses taken by the student. In light of these findings, it seems time that a test designed to fulfill the needs of the technical institutes should be developed. Dr. A. P. Johnson, of the Educational Testing Service, Princeton, New Jersey, has offered to work with a committee of this section in the development of a test which could be administered and scored locally for this purpose.

Karl O. Werwath, Milwaukee School of Engineering, chair of the Curriculum Development Committee, urged [that] schools having ECPD-accredited curricula establish institute curricula branches of the American Society for Engineering Education, and that through regular meetings of such branches, which include faculty men on their respective campuses, that a study be given to the educational and administrative problems in technical institute-type of curricula, so as to develop more professional literature for the field.

H. P. Adams, Oklahoma A & M, chairman of the Teacher Training Committee, urged the development of teacher training programs in at least four geographically distributed universities for the development of technical institute teachers.

Jeanne Miller, chair of the Technical Institute Studies Committee, reported that the technical institute guidance pamphlet has undergone a third revision by members of the division and there are sufficient orders to run a first edition of 50,000 copies.

June 1954   University of Illinois   Urbana
TID Business Meeting

Karl O. Werwath reports that during the past five years the Curriculum Development Committee has been engaged in making quantitative and qualitative analyses of the technical institute-type curricula, as accredited by the Engineers' Council for Professional Development. Some 90 courses of the technical institute-type in about 30 schools have been analyzed.

A. Ray Sims, University of Houston, reports that the Relations with Professional Societies Committee has found that, from a survey, each school has been actively engaged in building a better understanding of the place of the technical institute in higher education in its own local community. Some of the methods used in doing this are listed below:

1. Membership of the faculty in various professional organizations

2. Organization of student chapters of professional societies on campus

3. Serving as hosts to meetings of professional and industrial groups

4. Presentation of papers or talks before professional groups

5. Publication of papers in professional journals

6. The holding of career days

7. The establishment of central placement facilities, where recruiting teams and graduating students can meet and interview each other.

Don C. Metz, University of Dayton, reported for the Technical Institute Manpower Studies Committee on the enrollments and estimated number of engineering technician graduates for 1953-54.

	Enrollment		Graduates	
Type of Institution	Full-Time	Part-Time	Full-Time	Part-Time
State and municipal	4,400	915	1,346	298
Privately endowed	2,185	6,376	1,316	1,055
Extension, divisions of colleges & universities	2,480	8,310	415	945
Proprietary	4,797	1,900	2,242	408
YMCA Schools	48	439	12	46
	13,910	17,940	5,331	2,742

Jeanne Miller reported that the Technical Institute Studies Committee's "Literature Significant to Education of the Technical Institute Type," the 60-page annotated bibliography priced at $1 a copy, was published in November 1953. McGraw-Hill handled the composition and Capitol Radio Engineering Institute printed and bound the 1,500 copies. Secretary-Treasurer K. O. Werwath, who is handling distribution of the bibliography, mailed copies to 291 heads of teacher training schools. "The Engineering Technician," a 16-page guidance booklet priced at five cents a copy, was released in February 1954 through the McGraw-Hill Book Company. Out of a total printing of 50,000 copies, only 5,000 remain in stock on June 10, 1954. McGraw-Hill paid production and printing costs and is crediting payments against those costs. Arrangements are being made to reprint the booklet.

October 1954 Hotel Alms Cincinnati
Mid-Year Business Meeting of the Committee of 21 and TID Committee Chairmen

C. S. Jones, chairman of the Committee of 21, indicated that there was a need for development of a study on technical institute-type of education, bringing up to date the "Study of Technical Institutes," which was published by the Society for the Promotion of Engineering Education in February 1931. He appointed a committee to review this project and make recommendations at the next annual meeting.

The manner of expanding the treasury was reviewed by the chairman in his introductory remarks and he suggested that the members of the Committee of 21 voluntarily donate $10 each to establish a working balance.

A recommendation was made by Walter M. Hartung, Academy of Aeronautics, chairman of the Cooperation with Government Agencies Committee, to contact the U.S. Department of Labor regarding and including more acceptable vocational objective titles in their *Dictionary of Occupational Titles* for technical institute courses to aid veterans in filing for educational benefits.

June 1955 Pennsylvania State University State College
TID Annual Business Meeting

TID chairman, Charles S. Jones, reported that at the 1953 annual meeting of the society held at the University of Florida, a study conducted by the Technical Institute Division indicated a great lack of information, in fact, much "misinformation," concerning technical institutes. Consequently, the division decided that one of its main objectives would be to familiarize parents, prospective students, educators, the engineering profession, and industry in general with the purposes and names of the technical institutes, their place in the educational system, and the value of their graduates to industry. A pamphlet, "The Engineering Technician," showing the place of the technician in the engineering team, was published and is now in its third printing, and more than 65,000 copies have been distributed. In addition, numerous papers and articles have appeared in various periodicals and newspapers.

At the 1954 meeting at the University of Illinois, the General Council of the American Society for Engineering Education approved a resolution presented by the Technical Institute Division, authorizing the organization and development of a broad study of the technical institute field to be financed with the aid of foundation funds. During the year, this project has been organized with recommendations as to personnel and is now ready for presentation. A preliminary grant of $6,000 has been offered by Mr. Arthur Williston, formerly of Pratt Institute and Wentworth Institute, subject to the approval of the council, for the preparation of materials, curricula, and planning for the technical institute operation. In the meantime, shortage of engineers, great expansion of technology in many and varied fields, including the military, and broad technical programs in foreign countries have focused nation-wide attention on the importance of technically-trained personnel and the necessity for their proper use.

The activity of the Completion Credentials Committee was initiated under the direction of Mr. Kenneth L. Holderman of Penn State. Three years ago, the Technical Institute Division acted favorably upon the recommendation of the committee that the term "engineering technician" be used when referring to graduates of technical institutes and that a designated "associate degree" be used as the official award to those graduates. This year, the Completion Credentials Committee continued a study of the institutes as to their usage of the above terms. There are 92 institutes on the mailing list with a response from 75. Of those 75, a total of 38 are now awarding a designated associate degree with approximately the same number using the term "engineering technician." This represents an increase of five over the number reported last year. Two of these are this year awarding the degree for the first time, and the other three had not replied by the time of the report last year. Of the remainder not awarding the degree, there are 17 which have the matter under consideration for the future.

The Cooperation with Government Agencies Committee chairman, Walter M. Hartung, has made considerable contacts with the Departments of the Air Force and Army for the purpose of having them correct their regulations, so that students registered in technical institutes would be able to obtain their discharge up to 90 days early in order to meet scheduled starting days. The Cooperation with Government Agencies Committee has also had several meetings with members of the U.S. Department of Labor concerning the fact that the *Dictionary of Occupational Titles* did not include many of the objectives to which technical institutes graduated, or, in some cases, had titles which only partially applied to these technical institute graduates. An agreement was made between the editors of the *Dictionary* and your committee members which should result in corrections of these deficiencies. The Cooperation with Government Agencies Committee has made an attempt to keep abreast with new developments in the U.S. Office of Education. The director of the White House Conference on Education, Mr. Clint Pace, has been acquainted with the activities of the technical institutes, and your committee has offered its cooperation in his work. Contact has been maintained with the National Conference on Higher Education, particularly involving their activities concerning the draft of a bill to provide regulatory licensure of schools conducted for profit in the various states. Throughout all negotiations with government agencies, your committee encountered a very friendly and understanding attitude. A simple misunderstanding seems to exist in all cases where the technical institute has not been given proper treatment. That misunderstanding is not because proper information is not available, but rather that most of the people contacted had no idea that educational information concerning technical institutes existed. The work of your committee has been one of public relations, explaining to people that the technical institute program was a part of higher education and a very important part. It is recommended that renewed effort be made by all of the committees and all of the schools in the direction of publicizing the technical institute program, in determining a logical means of defining the program and in obtaining full recognition for the program.

In November of 1949, the current Curriculum Development Committee chairman, K. O. Werwath, issued the first of seven reports which analyzed the content of technical institute-type courses as accredited by the Engineers' Council for Professional Development. These studies were accomplished by teams of undergraduate research workers since then, with the last report completed during 1954. The original studies are in bound volumes and available through the registrar at the Milwaukee School of Engineering.

The Committee on the Place of General Studies in the Technical Institute Program chairman, H. Russell Beatty, was authorized to conduct a survey of the technical institutes to discover techniques used to accomplish general education objectives through both curricula and extracurricular activities. A questionnaire was prepared which listed 26 major general studies objectives, and these were further subdivided into 66 objectives. A factual study, indicating the attainment of general education objectives by an institution, is practically impossible. We have no real yardsticks for true evaluation at the present, but it is important that we think about our objectives and evaluate our accomplishments, so as not to become self-satisfied, even though the evaluation is based on opinion rather than a real measurement. Questionnaires were sent to all of the 27 technical institutes with ECPD-accredited programs and to additional technical institutes. The purpose of this study is to help those interested to discover techniques that may be used to broaden technical institute education to some degree.

Mr. Rimboi of the Relations with Industry Committee conducted a survey to "determine the possible use and feasibility of graduates as a point of contact between the graduate's school and the employer."

The 1954 report of the Relations with Professional Societies Committee, A. Ray Sims, chairman, pointed out many of the things that were being done by technical institutes to improve their relations with professional societies, industrial organizations, high schools, and the public in general. The past year has seen a continuation of progress made in this direction, but, frankly, we have failed in our major objective of gaining student membership rights in the AIEE for technical institute students enrolled in ECPD-accredited electrical curricula. "I am inclined to believe that a great many years will elapse before the original founding societies will open their doors to technical institutes. Even our own ASEE was in existence many years before the Technical Institute Division was created, and the ASEE is an infant when compared to some of the professional societies."

The Technical Institute Manpower Studies Committee chairman, Don C. Metz, sent questionnaires to some 207 institutions which records indicate have been offering training for engineering technicians. One hundred forty-three institutions furnished replies for both surveys.

The editor, Jeanne Miller, reported that, to date, two special issues of *Technical Education News* have been published, covering the 1953 meeting at the University of Florida and the 1954 meeting at the University of Illinois. We mailed 21,500 copies of each. About 16,500 went to the regular mailing list for *TEN*, which includes faculty members of technical institutes, personnel in military training programs, representatives of industry who are engaged in training, and a special list of guidance counselors, personnel directors, etc. McGraw-Hill's

College Department also mailed about 5,000 copies to engineering deans, presidents and deans of colleges and universities, directors of extension programs, and junior college faculties. Following the regular mailing, many requests were filled for additional copies for use in workshops, for distribution to boards of trustees, for faculty meetings, etc. Copies of both issues are still available.

The Study the Fund Committee chairman, E. E. Booher, has completed the proposed final draft of a request to Carnegie Foundation for a research grant for a survey of national needs and educational services at the engineering technician level and is in communication with the foundation for the granting of the funds.

October 1955   Ryerson Institute of Technology   Toronto National Committee of 21, Annual Mid-Winter Meeting

Chairman Karl O. Werwath, Milwaukee School of Engineering, reported that he was planning a monthly newsletter to go to all interested persons who had attended the past meetings of the division.

The Technical Institute Manpower Studies Committee chairman, Don Metz, submitted a preliminary report of the third "Annual Survey of Engineering Technicians" covering enrollments and estimated graduates for 1955-56. It is estimated that some 175 individual institutions present training for engineering and industrial technicians, frequently leading to the associate degree in the fields of engineering and science. Included are technical institutes, junior colleges, and divisions of colleges and universities. Last year, according to the study, there were 32,156 full-time students enrolled in curricula of the technical institute-type. This fall, that number has risen to 37,000, an increase of 15%. Total enrollments last year were 60,000, and this year the figure is 65,500, an overall increase of 10%.

It was recommended and approved that individual memberships be solicited for the division, the cost of the membership to be promoted by the division newsletter, a special brochure, and a membership card which is to be developed. It is also suggested that each institute appoint a membership chairman for the individual school. Kenneth R. Miller is chairman of the Membership Committee.

L. V. Johnson, chairman of the Curriculum Development Committee, pointed out that it had been indicated that 1,000 technicians are needed in the field of illumination, which might be a new area for curriculum development in certain schools.

The business meeting concluded, the members of the National Committee toured the Ryerson Institute of Technology and were photographed in the photographic department of the institute.

June 1956 Iowa State College Ames
TID Business Meeting

The editor, Jeanne Miller, reported 80,000 [copies of] "The Engineering Technician" distributed, with 23,000 on hand. Study fund granted $38,000 by the Carnegie Foundation. The U.S. Office of Education was asked to undertake a survey of ET enrollment and graduates similar to their engineering survey.

October 1956 Detroit, Michigan
National Committee of 21 Mid-Year Meeting

The editor reported approximately 500 copies of "Literature Significant to Technical Institute Education" were transferred to Ken Miller, Werwath, and Metz for distribution and sale. Special issues of *TEN* covering the annual meeting at Iowa State were mailed October 16, 18,400 regular and 5,000 to deans of engineering, instructors, librarians, extension personnel, colleges, universities, and junior colleges. Manpower Committee — U.S. Office of Education now to gather statistics.

Relations with Government Agencies Committee — DOT titles supplement for technicians being prepared. Civil Service Commission preparing new regulations to cover technicians.

June 1957 Cornell University Ithaca
Annual Meeting, National Committee, TID

After much discussion, a motion to hold the mid-year meeting in the Virgin Islands was lost.

June 1957 Cornell University Ithaca
TID Annual Meeting

These minutes will not be distributed to the membership inasmuch as complete reports from transcription will be issued in a special edition of the *Technical Education News*, volume 17, number 1.

Editor reported the sale of "The Engineering Technician" had exhausted supply, with the sales totalling over 93,000 copies.

The Cooperation with Government Agencies Committee reported that progress on the supplement to the *Dictionary of Occupational Titles* covering engineering technicians is moving along satisfactorily.

The Membership Committee reported 219 members. It recommended that dues be increased from $2 per year to $4 per year; motion carried.

E. E. Booher reported that the National Study Committee hopes that the greatest possible use will be made of the results of the work by Ross Henninger and his many helpers. Ross Henninger reported that the study was solvent and progressing on time.

October 1957 McGraw-Hill Book Company New York
National Committee, TID

The attendance of 29 was the largest to date for a mid-year meeting. The honored guests included W. L. Collins, secretary of ASEE; Bill Cavanaugh, executive secretary, Engineering Manpower Commission of Engineers' Joint Council, New York City; and Lloyd Slater, director of the Foundation for Instrumentation Education and Research, among others.

Russ Beatty strongly supported the need for a new edition of "The Engineering Technician" and pointed out that the National Science Teachers Association offered to send out 12,000 copies of "The Engineering Technician" to high schools, provided they are furnished free and $480 be provided for postage.

Secretary Johnson reported that prior to the Cornell meeting, one officer of the division served as both secretary and treasurer. However, because of the growth of the division, it was voted at Cornell to add another officer so that the division would have both a secretary and a treasurer.

Relations with Educational Organizations Committee chairman, Karl Werwath, quoted the last paragraph from a memorandum prepared for the National Commission on Accrediting. Mr. Werwath's quotation is as follows:

> *The National Commission on Accrediting is urgently requested to help make possible accrediting procedures through regional accrediting agencies so that such independent, nonprofit endowed technical schools may apply on the basis of standards of this type of higher education, just as universities and colleges offer other similar services. Each institution serves specific, well-defined objectives. Such accrediting procedure, it is requested, should then be based on the manner in which the individual institution meets its stated objectives.*

This memorandum was distributed through the National Commission on Accrediting to the regional associations and has resulted in action within the regional accrediting agencies: "Dr. Selden, at the National Commission, reports that three regional associations (Western, Mid-Atlantic, and New England), now consider technical institutes for accreditation; that the North Western, North Central, and Southern accreditation associations have appointed committees to review the general policy towards institutions not now accredited, and that at the fall meeting of regional associations this topic will undoubtedly be on the agenda."

June 1958   University of California   Berkeley TID Business Meeting

Chairman K. L. Holderman pointed out that the work of the Technical Institute Division is done by the committees of the division. The chairman and sub-chairman of the division can only channel the work of the division. The committee members themselves must do the work. He was glad to report that in a survey he had received over 100 replies (35% of the membership of the division) from members who are definitely interested in working on the various committees. This excellent response could provide new blood to regenerate committees and to provide information and articles for *TEN* and the division newsletter.

TEN editor, Jeanne Miller, reported that two publications were released during the year. One was a special issue of *Technical Education*

News covering the annual meeting held at Cornell University, June 1957. About 28,000 copies were distributed, and a stock is still available if anyone wishes copies for meetings, etc. Copies of three previous specials, covering the Florida meeting, 1953; the Illinois meeting, 1954; and the Penn State meeting, 1955, are still available. The 1956 special, covering the Iowa State meeting, is out of print. The second publication released was the revised edition of "The Engineering Technician." A motion was made and unanimously approved to give Miss Miller a vote of thanks for her fine service to the division.

Membership Committee co-chairman, Ken Miller, reported a total of 32 institutional memberships and 7 others pending.

Curriculum Development Committee chairman, H. H. Kerr, reported that European curricula vary with the country and the purpose of the school, but seem to be more specialized than those in the United States. England has a part-time and cooperative system leading to the Higher National Certificate. The German Technicums require industrial experience as a prerequisite but offer a greater range of theoretical subjects than the American technical institutions. The Russian Technicums provide a great deal of specialization. There is no uniformity in Russian curricula, but the training is very intense in a very narrow field. These examples show the trends of European technical schools.

It was moved by Ray Sims, and approved, that the Teacher Training and Recruitment Committee chairman, Maurice W. Roney, be authorized to draft a request to the National Science Foundation to call a meeting of 10 to 15 technical institute leaders to study means of establishing National Science Foundation summer institutes of technical institute teachers.

In order to create awareness of technical institutes, the Relations with Industry Committee chairman, Herb Hartly, recommends planning a joint meeting of members of the Technical Institute Division with members of industry — especially those employing engineering technicians. Such a meeting would provide an interchange of ideas and problems and benefit industry and technical institutes.

Ross Henninger received a unanimous vote of thanks and appreciation for his excellent work as project director of the ASEE "National Survey of Technical Institute Education."

Karl Werwath was given a unanimous vote of thanks for his excellent work on the President's Committee on Scientists and Engineers.

Chairman Holderman brought up the question of continuing the TID newsletter. After a discussion, it was found that the newsletter served a definite purpose, and it was voted that it should continue on a three issue per year basis from the chairman's office.

October 1958   Jefferson Hotel   St. Louis
Mid-Year Meeting, National Committee, TID

Editor Jeanne Miller reported that 75,000 of "The Engineering Technician" were printed before the annual meeting at Berkeley last June. Of these, 27,000 copies had been sold. She sent 400 copies to ASEE, and the Office of Education later bought 4,000.

Student Selection and Guidance Committee, J. J. Gillespie, chairman, raised two questions for consideration: (1) What were individual institutions doing about student selections through tests? and (2) Should an attempt be made to establish a testing program, national in scope, for technical institute student selections? Member institutes of ECPD will be asked to submit results of student selection methods through the use of standardized or locally-tailored tests. It was generally concluded by the National Committee that the development of specific standardized tests for technical institutes should not be attempted by the Student Selection and Guidance Committee, but rather is a problem to be handled by professional specialists in the area of test construction.

The Curriculum Development Committee chairman, H. H. Kerr, pointed out the inroads the vocational education people are making in the field of technical institute education. He pointed out that the vocational group is 100 times stronger than the technical institute group, and that some vocational programs have higher standards than some so-called technical institutes. Dr. Kerr urged all to see that the curricula of their institutes were maintained at a very high caliber, high enough to assure the program to be of freshman and sophomore college level. He also urged that every effort be made to tie each curriculum in with its professional organization or industry. To distinguish technical institute graduates from the so-called technicians from vocational schools, Dr. Kerr suggested that the name "technologist" be used for technical institute graduates.

ECPD is considering the problem of evaluating technical institute education. *TID Newsletter*, volume 5, number 1, January 1959, reads in part as follows:

Evaluation of Technical Institute Education. A project that unquestionably will have its far-reaching effects on technical institute and engineering education has been discussed and initiated during the past several months. Through the interest and efforts of President H. Russell Beatty, Dean Maurice Grainey, and other members serving on the ECPD Sub-committee on Technical Institutes, Engineers' Council took official action in October requesting the American Society for Engineering Education to make an evaluation of the technical institute education. This was done with the expectation that, from a thorough study, better accrediting criteria (both quantitative and qualitative) may be established.

Teacher Training and Recruitment Committee chairman, M. W. Roney, reported two significant summer programs conducted during June and July of 1958 for technical institute teachers. A training program under joint sponsorship of the American Society for Engineering Education and the Atomic Energy Commission was conducted by Pennsylvania State University and Argonne National Laboratory. The Foundation for Instrumentation Education and Research, in cooperation with the National Science Foundation, sponsored an instrumentation technology program with the Case Institute of Technology in Cleveland.

Relations with Educational Organizations Committee chairman, K. O. Werwath, then mentioned that there are many questions developing about the future of technical institute education: Should technical institutes be integrated as a part of the engineering profession? Should the U.S. offer licensing to technical institute graduates? The plan for this would be an exam for the students and then an advanced exam for professional licensing. The answers to these questions must be found.

It has developed that the "Technical Institute Enrollment Survey" compiled by the U.S. Office of Education is considerably out-of-date when it is published. It was therefore moved, seconded, and passed that D. C. Metz be authorized to reinstate his annual "Survey of TI Enrollment and Graduates."

L. P. Went addressed the members present, and his remarks are summarized as follows:

The unionization of engineers and technicians is a problem I very definitely feel should be seriously discussed by this society. As you all no doubt know, the AF of L and CIO in their last national convention publicly announced their intention of increasing their ranks through the unionization of white-collared workers, meaning, of course, not only office workers but technical and engineering employees as well.

In addition to this, the Engineers and Scientists of America, an independent union of technical and engineering employees, has announced their intention of trying to create student units in the various technical and engineering schools. All types of unions are making overtures to the white-collared employee. Approximately half of the 189 unions reported in the Bureau of Labor Statistics Directory of National and International Unions in the U.S. *(1957) are interested in this group.*

In many instances, the matter of adequate representation is of little importance — all that seems to matter is that the white-collared employees increase the size of the unions, add to the financial stability, and offset the loss of production workers because of automation.

Some of the major unions active in their unionization of white-collared employees and their estimated membership are as follows:

Engineers and Scientists of America	*15,000*
Engineering and Scientists Guild	*18,000*
American Federation of Technical Engineering	*15,000*
United Automobile Workers (office, engineer, and technical groups)	*100,000*
United Steelworkers of America (office and technical workers unit)	*40,000*
International Union of Electrical Workers (professional, technical, and clerical groups)	*3,000*
Office Employees International Union	*60,000*
International Brotherhood of Teamsters	*10,000*
Communication Workers of America	*125,000*

It is "open season" for the unions in their attempt to unionize all non-supervisory, salaried employees, and all technical and engineering employees are fair game. Their target is some 18 million unorganized white-collared employees, of which approximately 600,000 are engineers and technicians.*

* Steel Magazine, *July 1957*

The seriousness of this problem is brought out with one look at the attached appendix [to the original minutes] *which lists the unions whose membership is composed entirely or almost entirely of technical and engineering employees.*

Technicians and technical employees as a whole feel that automation is beneficial to them in that it will create more jobs for technically-trained persons. However, the unions are using this one word, "automation," as a "boogie-man" to scare technically-trained employees into unions.

I feel that it behooves each of us who directs a technical institute of engineering schools to teach something of management's side of the labor-management movement in order that graduates are not immediately thrown to the "wolves" upon graduation.

May 18, 1959 Memo to Members of National Committee, TID

A lot must be accomplished during our annual business meeting in Pittsburgh, so in the interest of time would you please type your committee report and give me three copies. The employment of a stenographer at the mid-year meeting proved impractical. Therefore, I must depend on my own feeble efforts — and that ain't good. — L. V. Johnson, secretary, TID

June 1960 Purdue University West Lafayette
TID Business Meeting

Ken Holderman, representative on General Council, stated that the "Survey of Technical Institute Education" has been successfully terminated. The "Williston" provision of $5,000 had not been used and some future use in the way of an award might be developed. Secretary Collins has advised the representative of a need for an improved communication between the TI Division and the council officers.

Jeanne Miller, editor, reported that the major communications and committee reports had been printed in *Technical Education News*.

Howard Kerr, chairman of the Curriculum Development Committee, stated that the work of the committee had been mostly confined to a study of technical institute mathematics. Curriculum changes had been few, and the committee work was largely on review and consolidation. A survey of course titles indicated a need for more unity.

A. P. Coleman, chairman of the General Education Committee, presented the report in the form of a proposal for a course of study in general education at technical institutes.

The division had 299 members in October 1952 and has grown slowly but steadily to a membership of 391, as of June 1960.

Don Metz reported on the fifth "Survey of Engineering Technician Enrollments and Graduates." The major problem is to identify bonafide technical institute courses and to verify the replies.

Karl Werwath, chairman of the Committee on Relations with Educational Organizations, reported that the committee had been working with the U.S. Office of Education and that suggested changes in Title VIII of the National Defense Education Act would be considered in new business.

Walter Hartung, chairman of the Relations with Government Agencies Committee, commended Ken Brunner for his fine efforts to further understanding of technical institute education in the various government agencies.

H. E. McCallick, chairman of the Relations with Industry Committee, stated that the committee has been working with industry to extend information to the high school counselors. Excellent cooperation has been experienced during the year.

Thor Trolsen, chair of the Relations with Professional Societies Committee, reported that the Engineers' Joint Council and the National Society of Professional Engineers had set up committees to study their relations with engineering technicians. The NSPE is considering the formation of an institute to evaluate the training of engineering technicians. This institute would have no legal status, but it could help in gaining status and recognition for the engineering technician. ASEE should cooperate with these organizations. Karl Werwath supplemented the report, stating that the mechanics of the institute were now being worked out by NSPE and some published matter should be available by the end of the year.

Karl Werwath, chairman of the *ad hoc* Committee on Reorganization of the Division, reported that a general need had been recognized for some type of administrators' group to expand the activities in technical institute education. The committee recommended some type of institutional membership classification for schools with ECPD accreditation.

June 1961 University of Kentucky Lexington
TID Annual Meeting

Chairman A. Ray Sims asked the chairmen of committees to make brief reports at this meeting since a reasonable summary would be published in the special edition of *Technical Education News*.

K. O. Werwath, chair of the Relations with Educational Organizations Committee, reported that the teacher exchange program of the English Speaking Union may be expanded to include administrators. An informal discussion indicated an interest in student exchange as well as teachers. It was voted to explore the possibility of enlarging the scope of teacher exchange with West Germany.

Chairman Sims reviewed the work of the *ad hoc* Committee on Reorganization of the Division and the possible creation of the Technical Institute Council. This type of reorganization would create an administrative group and an academic group. The division function under the proposed reorganization would be to carry forward matters in the academic area, especially matters of interest to technical institute teachers. Chairman Sims emphasized the need for reorientation of the division in terms of the basic aims of ASEE. During the interim, the TI Administrative Council would act as a committee of the division, and nominations of officers should be made to facilitate the integration of the academic and administrative groups during a possible period of transition. [An attendance roster was attached to the minutes of this meeting.]

June 1962 Air Force Academy Colorado Springs
TID Business Meeting

V. E. Neeley reported on the NSPE Institute for Certification of Engineering Technicians. The institute was established in 1961 and some applications have been considered to date. Voluntary action is required for any individual desiring certification.

K. O. Werwath, chairman of the Relations with Educational Organizations, reported that 10 teachers had been sent to the United Kingdom on an equal exchange basis.

K. O. Werwath reported on the revised constitution of ASEE and the formation of the Technical Institute Council. The Technical Institute Administrative Council Committee would now become the council, and, as such, function independently of the division. Mr. Werwath commented on the depth of understanding of the ASEE officials and expressed gratitude for their support in the establishment of the council. At the final meeting of the TIAC on June 19, 1962, certain recommendations were made for the division of responsibilities between the division and the council.

Dean Maurice Grainey, University of Dayton, reported that the late Mr. Arthur Williston gave ASEE securities to finance the writing of a history of the technical institute movement in the United States. The death of Mr. Williston caused complications and no decision had been reached regarding the method to carry out the intent of the donor. The present committee now proposes the formation of a special committee to be called the Arthur Williston Award Committee. Its function is to make recommendations for annual awards for literature on technical institute education.

C. C. Tyrrell reported that the "Evaluation of Technical Institute Education Study" had been edited and it was expected that printing would be done in the spring of 1963. [An attendance roster listing 71 individuals was attached to the minutes.]

June 1963   University of Pennsylvania   Philadelphia TID Annual Meeting

Chairman H. H. Kerr explained the reorganization of the division and its relation to the Technical Institute Council.

Karl Werwath, in reporting on the council, stated that its formation was an historic event and that TIC was now one of three major councils within the organizational framework of ASEE. This was a fitting culmination to 20 years of dedicated work by the division. The new image of the council should be helpful in promoting cooperative solution within ASEE of the possible extension of the two-year programs to three- or four-year programs. Newsletters of council activities are planned for the coming year. It is the objective of the officers of the council to get all people working in the field of each one's special interest. It was generally felt that the council was now a real part of the parent organization, ASEE. Due to the assumption by the Technical Institute Council of many of the functions of the Committee of 21, the By-Laws Committee proposed a two-year transition in phasing out the Committee of 21, after which there shall be an executive committee of seven members. The committee also suggested that the name of the division, originally submitted as the Engineering Technology Division, be retained as the Technical Institute Division. A prolonged discussion followed in which the history of the Committee of 21 was reviewed and the purpose of the division was examined in light of being called the Technical Institute Division or the Engineering Technology Division. The motion on the floor was amended to use the name of Technical Institute Division and the purpose to be the promotion and development of technical institute education. It was voted 36 to 25 to adopt the by-laws, with these exceptions, as submitted by the committee.

Harry Roundtree, Program Committee chairman, stated that the success of the 1962 program on curricula prompted the committee to continue the idea of acquainting teachers with the various curricula. [An attendance roster was attached to minutes of these meetings.]

December 1963 Milwaukee, Wisconsin
TID Executive Committee Meeting

Walt Shaw suggested that TID request a page regularly in the new ASEE journal to tell our story.

A national study for teacher training was proposed to formulate guidelines for curricula, which would provide satisfactory technical institute teachers.

June 1965 Illinois Institute of Technology Chicago
TID Business Meeting

Vice-chairman Richard Ungrodt reported that Ed Fleckenstein, the technical institute historian, is well along on the history of the division, and he again urged the cooperation of the membership in supplying Mr. Fleckenstein with any information he needed.

Richard Ungrodt reported that the Technical Institute Council voted unanimously to change its name from the Technical Institute Council to the Engineering Technology Administrative Council. Winston Purvine, of Oregon Tech, moved that the Technical Institute Division name be changed to Engineering Technology Division. A representative from the Queensboro College of New York then asked if this recommendation implied exclusion of institutions engaged in training "industrial technicians." Purvine rose to the point and evolved logically that they should be included and it was intended that they would be. The motion then passed unanimously.

Chairman Ungrodt called attention to the recent article written by ASEE President Easton, who suggested that cross-pollination between the various divisions of the society is very desirable and could be effected by integration of program planning. Walter Hartung pointed out that this is the way "to join the society," especially in the case of section meetings. The TID has been concerned for some time because it has been virtually excluded from section meetings. He pointed out that it would be far more to our advantage to join in the existing regional meetings than to try to set up separate meetings for the TID at the same time. At this point, G. Ross Henninger, arose to support wholeheartedly the comments of Hartung, and moved that an *ad hoc* committee be established to correlate our activities with those of the various sections. This was seconded and passed unanimously.

Ungrodt strongly recommended that every person present submit articles to the ASEE journal for possible publishing. He also recommended that those present encourage others of their various organizations to do likewise.

Ungrodt also asked for more active participation from the membership in deciding and manning the programs for the next year.

The council passed unanimously a Werwath resolution to the ASEE board that "A special issue of the *Journal of Engineering Education* is needed to bring to the American public at this crucial time authentic information on engineering technology, especially as to occupational guidance, career counseling, industrial needs, employment opportunities, educational programs, accrediting of curricula, and other related information."

Robert H. Thompson, chairman of the Membership Committee, reported that the outdated TID membership pamphlet was revised by Walt Shaw.

June 1966   Washington State University   Pullman
TID Business Meeting

Mr. Mike Mazzola, chairman of the Section Activities Committee, gave a report on this committee.

Mr. Ed Fleckenstein, Temple University, historian, gave a report on the history project of TID.

Mr. Walter Hartung, chairman of the Technical Institute Council, advised that the name of the Technical Institute Council has been changed to Technical Institute Administrative Council (TIAC). Our relation with the engineering societies is better now than it has been in years.

1965-66   TID, Report of the Chair, R. J. Ungrodt

The new Section Activities Committee was established to aid in increasing the participation of TID members in section activities and to encourage other section members to take an interest in TID activities. Future plans will include at least one committee member from each section, who will act as direct liaison with the section committees or the institution responsible for organizing the programs of the various section meetings.

Edward L. Fleckenstein, historian, reported that the writing of the TIC/TID history was begun through the formulation of a definite plan this year. The chairman of this committee, working with two additional men, has compiled some 250 pages of history on various schools. The research on this history has been both difficult and time-consuming. Further information is being sought covering various subjects and events of importance to the development of the technical institute movement and the Technical Institute Division of the American Society for Engineering Education.

The consensus of the *ad hoc* Self-Study Committee indicates the need for a continued and strengthened Technical Institute Division which would become active in all areas of the society, including local section participation.

October 1966   Brown Palace Hotel   Denver
TID Executive Committee Meeting

Section activities chairman, Mike Mazzola, is vitally interested in getting more TID participation in the section meetings and [in getting] attendance at the section meetings from the community colleges.

It was also discussed that the advent of the four-year ET programs might be an inducement for some of the community colleges to participate in ASEE, since they would be involved in transferring some of their students from the two-year to the four-year degree program.

Walter Shaw, representative to *Journal of Engineering Education*, reported that selected papers from the Pullman meeting are being edited by Robert Hayes from Southern Technical Institute, to be published in the November issue of the engineering journal. This issue is devoted solely to the technical institute.

Walter Shaw, chairman of the technician careers booklet, advised us that the first publication of this booklet was titled "The Engineering Technician," of which 500,000 copies were sold, and then it was changed into a booklet titled "Technician Careers," and that 100,000 of these were sold. At the present time supply is exhausted, and he has requests for more.

1966-67   TID, Report of the Chair, R. J. Ungrodt

Program Committee chairman, Robert Wear, has again developed an outstanding series of sessions for the annual meeting in East Lansing, Michigan. Emphasis is being placed on new developments, including workshops on the four-year engineering technology programs, as well as special sessions on innovations and new technologies.

Section Activities Committee chairman, Mike Mazzola, indicates increasing interest in section activities was evident throughout the year. A number of sections have had technical institute programs incorporated in the regional meetings.

The Membership Committee chairman, Robert Thompson, reported that five regional representatives supported the membership solicitation activities. A net increase of 263 members to a total of 1,346 was reported through April 1967. Since this committee was formed in 1964, they have effected a 77% increase in TID membership. The increase in membership was accomplished through the following procedure:

1. Each TID member was contacted and asked to encourage his associates to join ASEE and select TID as one of their divisional interests.

2. Letters went to all presidents and activities coordinators of the member institutions of TIAC requesting them to take an active part in the membership drive.

3. A letter went out to all directors of the technical division of each community college asking that he, his staff, and faculty join ASEE and select TID as a divisional interest.

4. And, finally, letters were sent to the administrative heads of each institution which had received "candidate for accreditation" status by ECPD. They were encouraged to urge their staff and faculty to become members of ASEE and TID.

June 1967   Michigan State University   Lansing TID Annual Business Meeting

Technical Institute Administrative Council president, Larry Johnson, reported that since ECPD accepted the ET committee report indicating that criteria for evaluation of bachelor degree programs were accepted, ECPD would therefore, in the future, be in a position to evaluate, for accreditation, bachelor programs in engineering technology. He also reported that the special issue of the *Journal of Engineering Education* was a success and that 66,000 copies had been issued, 26,000 going to high schools, and that there were approximately 8,000 copies on hand. He stated that, hereafter, there would be no special issue of the *Journal*, since the Board of Directors of ASEE determined that no one council be granted a special issue for the advancement of their academic area. President Johnson urged the members to submit papers on subjects of interest for publication in the *Journal of Engineering Education*.

Mr. Robert J. Wear, Academy of Aeronautics, reported that the English Speaking Union Technical Teachers Exchange Program was reinstated for 1967.

June 1968 University of California Los Angeles
TID Annual Business Meeting

TID newsletter editor, J. R. Martin, reported that three issues instead of four, as budgeted, were mailed.

Sectional activities coordinator, Mike Mazzola, reported that 30 schools had appointed school representatives to work with his committee to help start, plan, and organize more sectional activities. Mike praised Joe Aidala and Ernie Stone for their work with sections.

September 1968 New Orleans, Louisiana
TID Executive Committee Meeting

The organization of TID on ASEE sectional regions is progressing very well. Ernie Stone was commended for the good job he was doing.

June 1969 Pennsylvania State University State College
TID Annual Business Meeting

Membership chairman, J. Malone, reports there is an active representative in every section. Membership in TID is presently 1,667 and growing.

Section activities chairman, Ernie Stone, indicates progress is being made in some sections such as New England, Southeastern, Mid-Atlantic, and others.

June 1971 United States Naval Academy Annapolis
TID Annual Business Meeting

Chairman Mazzola briefly reviewed some of the changes in the ASEE organization and discussed their effect upon the Technical Institute Division. The Technical Institute Division will now be under the Council for Technical Divisions. The four councils are the Council for Technical Divisions, the Council for Teaching and Learning, the Council for Graduate and Continuing Education, and the Council for Public and Specialized Services.

It was moved and seconded that the division name be changed to the Engineering Technology Division and the Executive Committee be empowered to take the necessary steps to recommend this name change to the board of directors. A supporting discussion followed, and the motion passed unanimously.

1971-72 ETD Annual Report, Ernest R. Weidhaas, Chair

Membership increased by 217 persons due in large part to an active membership campaign directed by Issac Morgulis, of Ryerson Technical Institute. Among his accomplishments, he designed and published a new version of our membership brochure.

The division wishes to report its extreme satisfaction with the new editor and editorial policy of the *Engineering Education* journal, which has permitted the number of articles on engineering technology education to be increased fourteen-fold. We feel this could not help but enhance the needed communication between engineering and engineering technology educators and administrators.

June 1972 Texas Tech University Lubbock
ETD Annual Business Meeting

Membership Committee chairman, Issac A. Morgulis, reported that the main thrust of the membership drive was carried out by a direct mailing to ETD members through the ETD sectional representatives. The membership drive, held during March 1972, was timed to reinforce the ASEE headquarters' mailings.

By-Laws Committee chairman, Frederick Berger, presented proposed by-laws that reflect the ETD name change and new mail election processes.

1972-73 ETD Annual Report, Ernest R. Weidhaas, Chair

Membership increased by 364 persons under the able leadership of Issac A. Morgulis, of Ryerson Polytechnic Institute. Our total membership now stands at 2,108, which constitutes one-sixth of the total ASEE membership. [See Appendix H for graphs of ETD membership.]

June 1973   Iowa State University   Ames
ETD Business Meeting

Membership chairman, Issac A. Morgulis, reported that the main thrust of the membership drive was carried out through ASEE headquarters' mailings and through personal contact at the section and member levels. As of June 11, the ETD was second in the divisional new membership contest. This year, 21.6% of the new ASEE members elected to join the ETD as compared to 16.6% in the previous year.

Historian Edward Fleckenstein sent word that at least part of the missing material had been found. Materials shipped in September 1971 containing the Technical Institute Division notes and transparencies were found in May 1973. Still missing are two boxes of manuscripts mailed at the same time.

Walter Hartung led a discussion on some of the key issues that the division will need to become involved with in the near future. These included proposed changes in the ASEE constitution and by-laws, issues and proposed changes in the dues structure, professional ethics, and improved relations between engineering and engineering technology.

June 1974   Rensselaer Polytechnic Institute   Troy
ETD Business Meeting

A short period of time was devoted to having Mr. Jesse Defore speak to the assembled members regarding writing articles and sharing articles with other members of the society through the various publications of the society. Mr. Defore solicited written material involving engineering technology from all interested parties.

Chairman Walter Thomas apologized for the fact that only one newsletter went out during the year, but funding was not available for the newsletters. It was only through the generosity of Mr. Winston Purvine and Oregon Institute of Technology that one newsletter was printed and mailed.

Membership chairman, Issac Morgulis, reported that the Engineering Technology Division had 25% of the new members in ASEE this year.

TCC chairman, Richard Ungrodt, spoke relative to the rapid growth of engineering technology and the Engineering Technology Division of ASEE. A brief report was given regarding reorganization of councils within ASEE. The members' involvement with all colleges, universities, etc. was discussed as well as becoming involved with writing worthwhile articles for society publications.

June 1975   Colorado State University   Fort Collins
ETD Annual Business Meeting

Bob Wear, TCC chairman, announced that a series of guidance brochures on the two-year and four-year engineering technology programs was being developed in conjunction with ECPD.

Chairman Walt Thomas expressed his concern that the Civil Service Commission excluded the BET graduates from engineering positions and that no appropriate solution was forthcoming.

June 1976   University of Tennessee   Knoxville
ETD Business Meeting

Dick D'Onofrio, Program Committee chairman, reported on the large increase in attendance for all the sessions sponsored this year.

Gerald Rath gave a report on the 1976 College Industry Education Conference.

Frank Gourley, TCC Publications Committee chairman, requested suggestions for the May 1977 issue of *Engineering Education*, which will be devoted to engineering technology.

June 1977   Grand Forks, North Dakota
ETD Business Meeting

Mike Mazzola, TCC chairman, announced REETS report assignments. Lead groups will develop a report, while supporting groups will critique and suggest changes.

Rolf Davey reported that the historian, Ed Fleckenstein, had passed away. Old records are unavailable. Jack Spille will attempt to assemble an archive collection to replace prior historical material and requested material to be sent to him.

Frank Gourley stated that the May 1978 issue of *Engineering Education* will be on engineering technology. Submit papers to him.

June 1978   University of British Columbia   Vancouver
ETD Business Meeting

Newsletter editor, Frank Gourley, reported that two issues of the newsletter were mailed during the year. The establishment of the section representatives appeared to work out very well, with good grassroots response from around the country.

Walt Carlson, TCC chairman, reported that the Engineering Technology Leadership Institute will in the future be more closely related to TCC.

Frank Gourley stated that the May 1979 issue of *Engineering Education* will be devoted to engineering technology. Relevant articles are needed and should be submitted to either Frank or the ASEE Publications Office.

June 1979   Louisiana State University   Baton Rouge
ETD Business Meeting

Ron Williams presented a resolution in support of the resolution passed by the Engineering Technology Committee of ECPD, which requests that the adjective "engineering" be used in the title of the committee in the newly proposed ECPD reorganization, which would be responsible for the accreditation of engineering technology programs. The segment of ECPD responsible for engineering technology programs as proposed would be called the "Technology Accreditation Committee," whereas the thrust of the resolution would be to label the committee as the "Engineering Technology Accrediting Committee." This resolution was passed unanimously.

Frank Gourley indicated that the ETD board had decided to make $500 available to support project activities (mini-grants) of individuals which would be of interest to all members of the Engineering Technology Division. Two- to three-page proposals should be submitted to the board, which will review the proposals and select the project or projects to be supported. This money was contributed from a surplus of the microprocessor short courses by Dick D'Onofrio. These short courses were conducted during the annual conference at Vancouver in 1978.

Durward Huffman of Nashville Tech announced that the third annual Engineering Technology Leadership Institute would be held in Nashville on October 15-17, 1979.

New chairman Gourley announced that section representatives were needed. He also announced that the membership brochure for ETD needs to be updated and that a volunteer is desired to assist in this.

January 1980 Ramada Inn Tucson ETD Board Meeting

Reports were presented on the three mini-grants which had been previously approved by a mail ballot of the board members. These were the "Engineering Technology Administrator Development Survey" by Ike Morgulis, the "National Listing of Institutions Offering Engineering Technology" by Jim Todd, and the "Engineering Technology" brochure by Harris Travis.

Ellison Smith, assistant to the dean, School of Engineering, Purdue University, addressed the meeting concerning the charge which he had been given by the ECPD office to develop some new engineering technology discipline brochures. He was given this charge since he has been editing ECPD guidance materials for several years. He asked for the assistance of the board in this regard, since he was not familiar with engineering technology programs. It was decided that Gourley would offer to ECPD and all professional societies and organizations the assistance of the TCC Publications Committee to review any engineering technology-related brochures which they developed. Harris Travis agreed to chair an *ad hoc* committee to review brochures.

Don Buchwald reported that Frank Gourley had prepared and he had printed 3,000 copies of a new brochure to promote membership in ETD and ASEE.

It was moved and carried that the support of mini-projects of benefit to engineering technology be continued using the BASS account funds.

June 1980   University of Massachusetts   Amherst ETD Business Meeting

Frank Gourley reported that the May 1981 issue of *Engineering Education* will again be devoted to engineering technology. He also announced that the monograph on "Engineering Technology Audio Visuals" was in the second printing.

Bob Reid encouraged the membership to express their views on the name of the presently labeled Technology Accreditation Commission to any of the ABET board members from other member professional societies whom they know.

January 1981   Dutch Inn   Lake Buena Vista ETD Executive Board Meeting

Tom Kanneman announced there would be an open forum meeting on the possible formation of an institute for engineering technology on Thursday.

Division chair Frank Gourley announced that he is seeking input from the membership on individuals that ETD might submit to ASEE as nominations for president of the society.

February 1982   Town & Country   San Diego ETD Executive Meeting

Jerry Rath reported that an *ad hoc* committee of the ETD, ETC, and ETLI has been formed to look into the state of goals and objectives.

Frank Gourley, as chairman of the Program of Work Committee, will lead this effort.

Harris Travis indicated that the engineering technology brochure is going to print soon, and that schools should get their orders to him at Southern Technical Institute.

Larry Wolf presented a suggestion that the division consider publishing a quarterly journal.

June 1982   Texas A & M University   College Station
ETD Business Meeting

Section representative coordinator, Ernie Stone, reported that the purpose of section representatives is to get ETD activities scheduled at section meetings. He also hopes to involve two-year college faculty in ETD activities at the local level.

Ron Williams indicated that the "Technology Education Comments" column has been established in *Engineering Education* as a place for ETD members to express their opinions to the ASEE membership, and he solicits input.

Larry Wolf reported preliminary results of an interest survey which indicated 90% in favor of an engineering technology journal and 80% approval of dues of $5 per year to support it. The Journal Study Committee will continue with additional development activities.

A policy was passed that "no position opening announcements were to be printed in the ETD newsletter."

February 1983   Lake Buena Vista, Florida
ETD Executive Board Meeting

It was noted, in order to publish the journal, the ET Division would have to establish a $5 membership dues per year, with each member of the division receiving a copy of the journal.

June 1983   Rochester Institute of Technology   Rochester ETD Executive Committee

A motion was made, seconded, and passed that the Program of Work Committee be terminated and that Frank Gourley be appointed to continue this work as the ETD goals and activities coordinator.

Larry Wolf commented on draft number three of the Engineering Technology Development Committee. He noted that there are a number of engineering technology programs co-existing with engineering curricula and that there is severe competition for resources. He expressed concern that the committee's recommendations would weaken his justification for funding, since he was competing with engineering programs for the same monies. There was a general concern for the activities of this committee and a suggestion that the committee move cautiously.

Chairman Rath reported that Ernest Stone had resigned as section activities coordinator. Fred Lewellyn has agreed to serve as section activities coordinator.

A motion was made, seconded, and passed to establish a committee chaired by the mini-grant coordinator to establish procedures to apply for mini-grants and to report the results.

June 1983   Rochester Institute of Technology   Rochester ETD Business Meeting

Eddie Hildreth, chairman of the Society Membership Committee, reported that there is still room for activity in recruiting new members. He encouraged all members to recruit individuals for membership in the division.

Larry Wolf reported that the vote for the Engineering Technology Division dues of $5 was 300 "yes" and 80 "no." As a result, the ETD board has approved proceeding with the *Journal of Engineering Technology*, with a target date of the first issue being the spring of 1984. Dean Wolf introduced Ken Merkel, who [will be] the first editor for the *Journal*.

Chairman Rath noted that an *ad hoc* Historical Center Committee had been established to develop guidelines and solicit proposals for the establishment of this center.

Frank Gourley, chairman of ETCC Publications Committee, reported that ASEE is contemplating a change in the publication of engineering technology articles. The May issue devoted to engineering technology education may be phased out and perhaps one article will be used in each issue. Also, the "Technology Education Comments" column by Ron Williams is being discontinued. Harris Travis noted that copies of the engineering technology brochure are still available.

January 1984   Sheraton Dallas Hotel   Dallas
ETD Business Meeting

Larry Wolf, production editor for the *Journal of Engineering Technology*, reported that the first issue will be printed in March and mailed to the membership of ETD. More than 30 papers were reviewed by peer reviewers and the editorial board. Approximately 10 will be published in the *Journal*.

Harris Travis announced that the supply of the engineering technology brochures had been exhausted and that in order to print additional copies, he needed an indication of the number of copies that would be ordered. Before reprinting the brochure, the salary information will be updated.

Frank Gourley discussed the role of the goals and activities coordinator in identifying potential projects, personnel for those projects, establishing a format for supporting progress, and determining specific goals for the division.

ETCC chairman, Gary Fraser, reported that the ETCC By-Laws Committee had submitted proposed by-laws which were being considered.

Sam Pritchett reported that the 1984 ETLI will be hosted by California State Polytechnic University in Pomona on October 21-24. The highlight of the program will be a one-day management workshop by the Management Education Team from Hughes Aircraft Company.

June 1984   Salt Lake City, Utah
ETD Executive Meeting

Chairman Tilmans congratulated the editorial board of the *Journal of Engineering Technology* for publishing a quality journal. All reports received indicated that the inaugural issue, published in March 1984, was well received. A question as to whether or not non-members of ASEE and ETD could publish in the *Journal of Engineering Technology* was raised. The answer was yes, non-members could publish in the *Journal*.

Lawrence Wolf reported that there were 34 engineering technology events planned for this 1984 annual conference.

A request for proposals of a site to house an engineering technology education historical center will be included in the fall newsletter.

Hank Stewart, of John Wiley Company, met with the Executive Committee and proposed an award for the engineering technology community to recognize excellence in engineering technology education. This award would recognize a department or institution for outstanding achievements in engineering technology. Since this award is for departments or institutions, the Executive Committee preferred to steer it to the Engineering Technology College Council as the most appropriate group to consider the award.

Frank Gourley suggested that the formation of special interest groups be considered to better serve the membership of ETD and that procedures be integrated into the goals and activities for the division to implement these special interest groups.

A report on trends in BET programs is planned, with a completion date set so the results will be ready for the 1986 CIEC. Some of the studies were done in 1978 and 1982, and by four ASEE zones. Frank Gourley asked that the next questionnaire be expanded to include two-year schools.

June 1984   Salt Lake City, Utah
ETD Business Meeting

Gary Fraser, chairman of ETCC, reported that the *ad hoc* Committee on Engineering Education Studies addressed the resolution prepared by the Engineering Technology Development Committee. This resolution proposed a study on the relative use of graduates in engineering and engineering technology, along with the meanings of degree designations and job titles for these graduates.

Jim Forman, membership chairman for the division, reported that the Engineering Technology Division currently has 1,701 members. It is the third largest division of the society.

Frank Gourley, chairman of the ETCC Publications Committee, called attention to the May issue of *Engineering Education*. He thanked Bill Schallert for his efforts in coordinating articles for publication in the May issue. Ron Williams will serve as coordinator for the coming year.

Frank Gourley has been asked to serve as the ETD goals and activities coordinator, and, as such, he is interested in the division establishing "special interest groups."

June 1985 Atlanta, Georgia
ETD Executive Committee Meeting

Ron Paré reported that there are 33 sessions sponsored by ETD at this meeting. Overall attendance is up 30% over the same time last year.

Durward Huffman announced that Jim Todd had completed the listing of 776 institutions offering engineering technology programs.

Jerry Rath made the motion that the ETD board recommend to its membership that Wentworth Institute house the historical center.

Larry Wolf presented, for ETD consideration, information on the proposed World Congress on Education in Engineering and Engineering Technology, to be held in Portland in June 1988, in tandem with the ASEE annual meeting.

Harris Travis stated that the engineering technology brochure was selling very well, so well that he had told the printer to go ahead with the next printing.

June 1985   Atlanta, Georgia
ETD Business Meeting

Bill Schallert reported that the section activity coordinators would meet for breakfast on Wednesday to discuss section activities. Fred Emshousen will take over as section activity coordinator for the next year.

Ray Sisson reported that the chairs of ETCC, ETLI, and ETD had formed a committee to locate a depository for the records of these three organizations. It was moved, seconded, and passed that Wentworth Institute of Technology be designated as the holder of the historical center.

Mike O'Hair asked that he be replaced as ETD historian. The new historian will be Ann Montgomery Smith, librarian at Wentworth Institute.

Bill Byers, of the Special Interest Groups Committee, reported that the first meeting of SIG's was held on Monday night.

ETD chairman, Tony Tilmans, discussed the World Conference on Applied Engineering and Engineering Technology which is planned to piggy-back with the 1988 ASEE annual meeting in Portland.

Frank Gourley, ETCC Publications Committee chairman, noted that the May issue of *Engineering Education* was once again devoted to engineering technology. Ron Williams is to be congratulated for his work as editorial coordinator for this issue. There was a discussion regarding an ETD membership directory. The estimated cost is $1 per member.

Harris Travis reported that the QEEP task force to devise a national framework for keeping engineering technology faculty up-to-date has met on three occasions. They have put together a tentative plan.

February 1986   Sheraton   New Orleans
ETD Executive Committee Meeting

Concern was voiced over expenses for the *Journal of Engineering Technology*, as they are much higher than income. Only 2,000 issues are printed, resulting in a high unit cost. There are, at present, only 60 libraries with paid subscriptions. After much discussion, Jim McBrayer made a motion that *JET* be sold in bulk to institutions for distribution to alumni and students for $5 per year, if this is at least production cost. This would be a trial for one year to determine the effectiveness of this approach.

H. T. Travis announced that sales are good for the "Engineering Technology Career" booklets. One school bought 5,000 copies. The brochures cost 50 cents each and there are a number left.

February 1986   Sheraton   New Orleans
ETD Business Meeting

George Timblin, program vice-chairman, noted that ETD will be sponsoring its first mini-plenary at an annual meeting. It will be on economic development in North Carolina.

A goals and activities committee has been formed by Chairman Travis, headed by Frank Gourley, to examine where ETD is relative to previously set goals and where ETD will go from here. The committee will meet at the June annual conference.

Fred Berger, executive director-secretary of Tau Alpha Pi, announced that there are over 130 chapters and that the honorary society is still growing.

June 1986   Cincinnati, Ohio
ETD Executive Committee Meeting

It was reported that George Timblin was recuperating from recent back surgery and that Frank Gourley had stepped in for the "fun part" of the job of ETD program chairman for this annual meeting. While Frank has "done good," the division thanks George for a job well done!

Support was again solicited for maximum ETD participation at the plenary session sponsored by the division.

Bill Schallert announced that a change in the printer had resulted in a reduction of approximately $400 in the production cost of the spring newsletter.

Mike O'Hair passed out information on bulk subscription rates for the *Journal of Engineering Technology*. It was noted that Southern Tech purchased approximately 100 copies and gave one to each graduate. If this practice were followed by a larger number of schools, this would have a positive effect on unit publishing costs.

Larry Wolf announced that the *JET* is now indexed by the *Engineering Index*. Approximately 60 institutional library subscriptions are now in force. There are still a limited number of complete sets of back issues available for purchase by institutions wishing to have a complete set.

Section activity coordinator, Fred Emshousen, handed out a preliminary "Manual for Section Activity Representatives," which included the purpose of the manual and the responsibilities of local section activity representatives, and asked that these be reviewed and comments be given to him at or before the business meeting.

The Goals and Activities Committee was formed at the CIEC meeting in New Orleans and is still in the planning stages.

Frank Gourley noted that the May issue of *Engineering Education* had been published. This issue emphasized engineering technology. Bill Welsh was the coordinator for this issue.

Monographs are supposed to bring in money for ETD through their sale at ASEE; therefore sales of these monographs are encouraged.

Frank Gourley handed out information on a possible Chicago-to-Reno train excursion for the 1987 ASEE annual conference.

June 1986   Cincinnati, Ohio
ETD Business Meeting

Larry Wolf announced that the editorial board of *JET* is interested in the possibility of electronic transfer of manuscripts to the printer.

An ETD membership directory is in the works and should be available in approximately one year.

Ann Montgomery Smith, historian, noted that it has been almost one year since the archives were transferred from Purdue to Wentworth. Negotiations are still ongoing with the University of Cincinnati for the transfer of their engineering technology holdings. Discussions are also continuing to better define what the archives should and should not hold.

Frank Gourley, ETCC Publications Committee chairman, noted that Bill Welsh was responsible for this year's May issue on engineering technology in *Engineering Education*. The membership voiced their desire to see the ASEE journal stay as it is and not go toward a research orientation.

Tony Tilmans announced a new award for engineering technology. The Wiley Award for excellence in engineering technology programs will first be awarded in 1987.

February 1987   Lake Buena Vista, Florida
ETD Executive Committee Meeting

Chairman Travis stated that the ET career booklet would have to be updated in the near future. Probably 70,000 to 100,000 copies have been sold overall.

Bob Buczynski, Penn State University, requested a mini-grant to survey 150 technical colleges of ASEE to determine what software is being used in ET programs.

Fred Emshousen, section activities coordinator, reported on activities to formalize section activities and submitted a copy of the "Section Activity Representative Manual."

February 1987   Lake Buena Vista, Florida
ETD Business Meeting

Ray Sisson, ETCC chairman, noted that by-laws changes were going to the membership for a vote. They will assist with the "infusion" of the Engineering Technology Leadership Institute into ETCC and the name change from ETCC to Engineering Technology Council.

Chairman Travis reported that the ETD Outstanding Service Award Committee had recommended that persons having made major contributions to the area of engineering technology education and who are retiring, going to industry, going into engineering education, etc., be given a plaque recognizing their efforts. Selection will be made by the chairpersons of ETD, ETC, and ETLI. Certificates of recognition will be given to all outgoing directors. Outgoing chairpersons of ETD and ETCC will receive plaques during the McGraw Award Banquet.

Bill Welsh is responsible for the May 1987 issue of *Engineering Education*. Fred Emshousen will be responsible for the May 1988 issue.

June 1987 Bally's Resort Hotel Reno
ETD Executive Committee Meeting

Historian Ann Montgomery Smith submitted a written report listing recent activities of the Historical Committee. Highlights include:

1. Plans for an "old-timers" project wherein a panel discussion by former McGraw awardees will be videotaped.
2. Collection of all McGraw Award speeches is underway.
3. The Don Metz papers, received from the University of Cincinnati in April 1987, are being examined and processed.
4. 1987-88 goals — gaps in the historical records are being identified. Members who have historical material and are in doubt as to its value are encouraged to send it to the archives at Wentworth Institute of Technology.

Bill Byers distributed a draft of a proposal for a professional development recognition program. The program would recognize engineering technology educators for professional development activities undertaken within a consecutive three-year period and would be renewable. The motion approving continued development of the proposal was approved.

June 1987 Bally's Resort Hotel Reno
ETD Business Meeting

Frank Gourley, chairman of the Goals and Activities Committee, commented on recent work to revise the ETD guidelines developed in 1983. The committee is also reviewing strategic planning activities. Bill Byers reported on a project to develop an award program recognizing continued professional development of engineering technology educators.

Fred Emshousen is editorial coordinator for the May 1988 issue of *Engineering Education*.

A Style Manual for Report Writing for students is available from Oregon Institute of Technology.

Fred Emshousen discussed new guidelines for section activities coordinators and identified the current coordinators.

June 1988   University of Portland   Portland
ETD Executive Committee Meeting

Bill Byers, ETD program chair, reported that there are 47 sessions scheduled that are sponsored or co-sponsored by ETD. Everything looks favorable at this point.

Frank Gourley and Fred Emshousen discussed the recent updating of the ETD guidelines.

Frank Gourley handed out draft copies of the ETD/ETC/ETLI brochure, which is being printed at Kansas Technical Institute. He has worked with these three groups to prepare the copy. Don Buchwald designed the new logos for ETC and ETLI, with Frank's input.

Jim Hales commented that in past years there has been little social interaction between the four divisions at CIEC. To help overcome this problem, Hales moved that ETD co-sponsor a hospitality suite at the 1990 CIEC.

June 1988   University of Portland   Portland
ETD Business Meeting

Frank Gourley, Society Publications Committee member, discussed a proposal to reduce the number of issues of the *Engineering Education* journal to six. In addition, there will be a directory of research, plus a directory of programs. The latter will include four-year programs, but not two-year programs. Stan Brodsky moved that the officers of ETD object strenuously to the ASEE board regarding this proposal; further, that the ASEE program directory should not exclude two-year programs.

June 1989   University of Nebraska   Lincoln
ETD Executive Board Meeting

Lyle McCurdy, ETD program chair, reported on the progress that the Peer Review Process Committee has made. The committee wanted to emphasize that paper submission does not guarantee printing in the proceedings. The timeline will be fully implemented in 1991. Frank

Gourley, Lyle McCurdy, and George Timblin will present a recommendation on how the process is to occur. A suggestion of having ETD proceedings was also presented.

Bob Buczynski presented an IBM academic software project proposal to produce high-quality software for students and faculty. Over one-third of 30 proposed programs are engineering technology-related. This software will be distributed at institutions in order to obtain more reviewers. There will be institutional subscriptions for the software.

June 1989   University of Nebraska   Lincoln
ETD Business Meeting

Journal of Engineering Technology editor, Mike O'Hair, announced that anyone wishing to publish in the *Journal* should forward the article, eight copies, and a 5 1/4" diskette to Howard Heiden. Everyone was asked to help maintain the high standing of the *Journal*.

Historian Ann Montgomery Smith stated that she needs more material and wants documentation of what is going on. She indicated that the archives were going to be easier to access due to the existence of a finding-aid and an OCLC listing. She also said that the Engineering Library's materials are being stored with the archives.

Goals and Activities Committee chairman, Frank Gourley, presented a proposed professional development recognition program for engineering technology educators. Frank also requested input from the membership on a way to document the activities of ETD.

Fred Emshousen announced that the pattern design for the Tau Alpha Pi key is complete.

February 1990   Grosvenor Hotel   Lake Buena Vista
ETD Executive Board Meeting

Tony Tilmans reported on the proposed 10-volume compendium on engineering technology. The volumes are ready to print; however, $10,000 is needed before printing. Prepaid orders will be sought as there will be no up-front money available to help publish the volumes.

February 1990   Grosvenor Hotel   Lake Buena Vista
ETD Business Meeting

Fred Emshousen reported that a form for consistency in the review process for proceedings articles has been developed. Mike O'Hair reported that the ETD approach is the best one that ASEE has received.

Durward Huffman reported that printing costs for the *Journal of Engineering Technology* are going up. The fall issue costs $16,000 versus $8,500 to $9,000 for previous issues. The main item causing this is typesetting. The spring '90 issue will be put on hold if the funding is not available. Changes and reductions will be made to keep production costs down. Some cost-cutting measures were suggested.

June 1990   Sheraton Centre Hotel   Toronto
ETD Executive Board Meeting

Journal of Engineering Technology editor, Durward Huffman, reported that there were 29 articles for print logged in last year: eight were printed in the fall '89 issue, seven in the spring '90 issue, and seven in the fall '90 issue. The fall '89 issue contained the first color article. Procedures are now available to do typesetting directly from the PC.

Don Buchwald presented a proposal for printing an ETD brochure.

Dave Hata reported on the status of the author-style guidelines for conference proceedings.

Dave Hata reported that the IEEE Computer Society will not contribute any funding to the Computer Engineering Technology Curriculum Guide Committee. However, ACM provided funding from a $60,000 grant for computer-related education. There may be up to $10,000 available for this work.

Steve Cheshier, ETC chairman, reported that at the business meeting, the members will be voting on replacing the Wiley Award with the Frederick J. Berger Award.

Ann Montgomery Smith reported that the archives are alive and well.

June 1990   Holiday Inn   Toronto
ETD Business Meeting

Jim Forman's membership report was not presented due to his being unable to get a list of new members from ASEE headquarters.

Frank Gourley reported that the *Directory of Engineering Technology Institutions and Programs* is finished. All data were printed, including accreditation. The *Directory* is available from ASEE headquarters.

Goals and Activities Committee chairman, Frank Gourley, reported that the ETD Executive Board recommended the establishment of a professional development recognition program. It will be run for a two-year trial period and then be re-evaluated. It is to be used by and for ETD members only to recognize their peers for outstanding professional development. George Timblin announced that an implementation committee will consist of Bill Byers (chair), Frank Gourley, and David Hata.

January 1991   Town & Country Resort Hotel   San Diego
ETD Business Meeting

Fred Emshousen distributed the preliminary program for the 1991 annual conference. There are presently 134 individual presenters for ETD.

Ann Montgomery Smith, historian, reported success with her request for materials which she presented at the last annual meeting. She received quite a few items and was able to fill some significant blanks. She is now looking for photos for the archives. She again asked all members to turn over their committee files to her rather than throwing them out, as there may be valuable material in them.

Frank Gourley reported that the Goals and Activities Committee is working on the goals for the division as they relate to the mission and vision statements.

Mike O'Hair, Centennial Committee chairman, reported that the committee plans to publish a history of ETD/ETC. Other tentative plans discussed ranged from a plenary session at the 1993 annual conference to a time capsule.

Steve Cheshier, ETC chair, reported that the ASEE Executive Board had adopted the Strategic Planning and Mission Statements with the five strategic priorities, including "defining a structure for engineering education for the twenty-first century."

Tony Tilmans, Kansas Technical Institute, reports the ETD 10-volume compendium continues to be a money-maker.

The ETD membership directory status is on hold until we receive an updated database from ASEE headquarters. The *Directory of Engineering Technology Institutions and Programs* is still available from ASEE headquarters.

The Ben Sparks Award is funded and will be awarded for the first time this year at ASME's annual meeting.

January 1992   Alexis Park Hotel   Las Vegas ETD Business Meeting

There was considerable discussion regarding the financial condition of the division, given the expenses of printing the newsletters and the *Journal of Engineering Technology*. Possible ways to help fund the *Journal* were discussed.

Earl Gottsman, chairman of the Nominations Committee, reported that 350 ballots were returned out of the 1,100 sent out. He commented that the process of sending out individual ballots should continue, as a much greater return occurs than if the ballot is placed as a tear-out page in the newsletter.

Gary Crossman, program chairman, reported that the ETD program to be held in Toledo, Ohio, is well developed with a total of 36 sessions. There are 19 technical sessions, 4 TAC/ABET sessions, 3 poster sessions, and 10 luncheon and/or business sessions.

It was moved and seconded to have the ETC Publications Committee study the advisability of publishing mini-grant reports as monographs and to recommend to the Executive Committee a procedure to accomplish the sales with the revenue returned to the ETD treasury.

Elliot Eisenberg reported that there were 24 articles submitted to the *Journal of Engineering Technology* for the spring issue and that seven were accepted for publication.

Steve Cheshier, ETC chair, reported that the ETLI is examining the possibility of moving to some other time of year or piggy-backing on the CIEC or the annual ASEE conference. Steve indicated that a committee has been established, consisting of three engineering and three engineering technology persons, to work jointly regarding the acceptance and implementation of the 11 principles of engineering technology.

June 1992   Radisson Hotel   Toledo
ETD Executive Committee Meeting

Gary Crossman, program chairman, reported that there were 124 presenters with 75% of the papers produced in the *Proceedings*.

June 1992   Radisson Hotel   Toledo
ETD Business Meeting

Membership chairman, Rolf Davey, reported that the ETD's membership was approaching 1,300. He challenged every member to recruit one new member this year. Fred Emshousen discussed the mechanism of ASEE's sending an application with no opportunity to join divisions on initial application.

Section activities coordinator, Al Avtgis, reported that the New England Section Conference held October 1991 included two presentations related to engineering technology. Richard Moore presented a paper and a workshop during the Pacific Northwest Section Conference. Gary Crossman chaired the Southeast Section Conference where there was considerable activity.

Warren Hill reported that the ETLI board will be recommending to ETC that the fall 1993 ETLI meeting be moved to coincide with the 1994 CIEC meeting and be run for the first two days of the CIEC in a workshop format. The proposal must be approved by both ETD and ETC.

Elliot Eisenberg reported that the *Journal of Engineering Technology* workshop run at the beginning of this conference attracted 20 people. He expressed concern about the small number of papers submitted. So far, 12 have been submitted for the fall issue, with only three accepted.

Part II: Excerpts from Council Meetings 1964-1993

1964-65 TIC Annual Report, Walter M. Hartung, Chair

A study was initiated to determine the feasibility of accepting membership into the Technical Institute Council from among institutions not having programs accredited by the Engineers' Council for Professional Development, but whose programs could well be considered equivalent to those having such accreditation. As a result of this initial study, it was the decision of the American Society for Engineering Education Board of Directors to consider Canadian schools, properly accredited by the Association of Professional Engineers and the Province Department of Education, acceptable as Technical Institute Council members upon the recommendation of the council. As a result of this action and with the approval of the Executive Committee of the Technical Institute Council, institutional membership has been offered to the Ryerson Institute of Technology.

The tenth "Survey of Engineering Technician Enrollments and Graduates" for the year 1964-65 reported significant increases in all statistics. A total of 177 institutions reported, 33 of which have at least one program accredited by the Engineers' Council for Professional Development. The total enrollment in all institutions, full-time and part-time, as of October, 1964, totaled 65,731 students.

A study was conducted among the member schools of the Technical Institute Council which would reflect the industrial, academic, and social status of its graduates. A cross section of graduates over a space of several years reported a high degree of success in present position titles. The survey revealed that a significant percentage of technical institute graduates have continued their education to the bachelor's degree and beyond. Engineering technology graduates reported a satisfactory record of participation in professional and community activities. This survey should prove useful in future projects by providing statistics indicating the success probability for graduates of the technical institute programs.

This year, four American teachers have been selected to participate in the technical institute teachers' exchange program sponsored by the English Speaking Union of the Commonwealth.

A most significant action on the part of the Technical Institute Council has been its cooperation with the new American Society of Certified Engineering Technicians.

Plans were made to publish an issue of the *Journal of Engineering Education* during the summer of 1965, devoted entirely to technical institute education and featuring, specifically, guidance.

Close cooperation continues between the Technical Institute Council and the Technical Institute Division.

February 1966   TIC Chair's Report, Walter M. Hartung

With major developments taking place in the area of engineering technology, such as a consideration of four-year programs, and re-identification of "institutes" to "colleges" and, in general, an expansion of activities, the Executive Committee of the Technical Institute Council considered it appropriate to prepare policy statements concerning the council's relationships to and within the ASEE. The Technical Institute Council desires to remain a contributing part of the ASEE and is committed to work constructively at all times for the benefit of the society. There is need for a close relationship between engineering technology and engineering from both the educational and industrial viewpoint.

The National Science Foundation has been asked to support "The Goals of Engineering Technology" study. Jesse Defore of Southern Technical Institute is the agreed-upon project director.

The Technical Institute Council is interested in publishing an eleventh issue of the *Journal* during the summer of 1966. The major purpose of this issue would be to present information concerning the Technical Institute Council program which would be of value for the guidance of high school students and of use to counselors. A letter from the editor of the *Journal* stated that a press run of 125,000 copies would cost approximately $25,000. The Technical Institute Council Committee for the Eleventh Issue plans to finance the publication as follows: member schools would take one-page advertisements at $600 each. Assuming 38 schools in this program participate, that would be $22,800. It is anticipated that as many as 40 companies would take one page of advertisement at $600 each, which would result in $24,000. The total anticipated income from these two sources would be $46,800. At the present time, approximately 12 schools indicate their financial support prior to any generalized approach to all schools. Furthermore, 13 companies have been approached on a trial basis, and out of these, 12 indicated that they would support the issue by advertising. Based upon the good probability of financial success of the issue, permission was requested from ASEE to proceed.

June 1969 Pennsylvania State University University Park
TIAC Business Meeting

ECPD Policy Committee chairman, Walter M. Hartung, pointed out that approximately 85% of the eligible engineering institutions were accredited by ECPD, but only 40% of the engineering technology schools were seeking ECPD accreditation. Criteria for evaluating four-year bachelor programs in engineering technology will be submitted to the National Commission on Accrediting with the request that ECPD be formally recognized as the agency for accrediting bachelor's programs in engineering technology.

Guidance publications in engineering technology are nearly completed and will be printed soon by ECPD. The Engineering Manpower Commission will cooperate in widespread distribution. Plans are now underway for a spectrum-type guidance piece which will compare and cover all aspects of the engineering profession; *i.e.*, engineering technology, professional engineering, and scientific endeavor.

TIAC Publications Committee chairman, Robert W. Hays, reported that the demand for "The Engineering Technician" pamphlet has been exceptionally heavy, and it is widely accepted as an outstanding piece of guidance literature.

June 1970 Ohio State University   Columbus
TIAC Meeting, Report of the Chair, Eugene W. Smith

The background of the "Engineering Technology Education Study" was detailed. The grant was made last summer by the National

Science Foundation and Dr. Jesse Defore was selected by the Projects Operating Unit of ASEE to organize the detailed study under the project director, Dr. L. E. Grinter of the University of Florida. An advisory committee was selected and several meetings have been held. A preliminary report on the study is scheduled today. The outcome of this study will be most significant to engineering technology education.

The Long-Range Planning Committee of ASEE and a special committee on reorganization of the society have been working on plans to change the society structure to meet more effectively the challenge of the future. The officers-elect of TIAC will have a most challenging responsibility to protect the integrity of the council in the final reorganization plan.

The institutional membership of the council has been increased during the past year, and it is a pleasure to welcome the representatives of the new institutions. We have a real opportunity to increase our affiliate membership, especially in the junior college area. It is important that we work closely with the public and private junior colleges which have engineering technology programs as well as parallel engineering programs.

June 1970 Ohio State University Columbus TIAC Business Meeting

Manpower Committee chairman, Don Metz, reported that the Engineering Manpower Committee of the Engineers' Joint Council has continued its generous support of engineering technology with many surveys parallel to those for engineering. Hopefully, EMC will continue with these data gatherings and excellent reports.

Relations with High Schools Committee chairman, H. B. Ellis, reports that probably the one thing that adversely affects our relations with high school students and their parents the most is the adverse connotation applied by most baccalaureate institutions to the "terminal" or "occupational oriented" title applied to two-year programs. This can be dissipated to some extent if it can be shown that opportunities do exist for continuation on to the sacred bachelor degree. He proposed that the committee compile a guide for students considering a two-year engineering technology program, which will give them in concise form the requirements and opportunities for continuing education at any institution that would welcome them. A third condition which adversely affects our relations with the students in high schools is the lack of understanding of what engineering technicians do in industry.

Relations with Professional Societies Committee chairman, Joseph J. Gershon, reported that IEEE has been particularly active and progressive in the area of education. Efforts to introduce innovative programs for members are obviously paying off, inasmuch as membership remains over 160,000.

The *ad hoc* Computer Software Committee chairman, C. L. Foster, reported that a final report by the Joint Committee of ASEE and ACM concerning development of a new curriculum titled "Computer Software Technician" had been made to both ASEE and ACM.

June 1971   TIAC Annual Report, W. D. Purvine, Chair

The outstanding event of the year is the society reorganization and, with it, the revised name — Technical College Council.

Suggested objectives of this council were received as follows: (1) to provide a forum for discussion of problems and exchange information pertaining to technical colleges; (2) to provide programs at meetings; (3) to represent and speak on behalf of technical colleges; (4) to improve the effectiveness of technical college administration. The suggested objectives are expected to be assigned to the Long-Range Planning Committee for consideration and an early report. As a special order of business, the 25,000th certificate to an engineering technician is to be presented by ICET on June 22, 1971. This landmark achievement by a cooperating organization is to be congratulated.

The ASEE-NSF "Engineering Technology Education Study" began as a goal of the TIAC's Special Projects Committee to develop an updating study. The Projects Operating Unit assisted in reaching the success of obtaining the NSF grant.

During the last two years, the study has been vigorously directed by L. E. Grinter. Assisted by M. R. Lohmann, Jesse J. Defore, and a working advisory committee, the study nears its end. Hundreds of engineering and engineering technology educators have responded. Many professional societies, industrial firms, and others made contributions. It is due to be a final report by October 1971 and to be presented at the annual ECPD meeting. Accreditation standards may then be devised from data presented in the study report. This is a very significant activity that is about to come to a successful completion.

June 1971   Annapolis, Maryland
TIAC Business Meeting

The Institute for the Certification of Engineering Technicians representative, C. L. Foster, reported that up to June 1971, a total of 632 current graduates of ECPD-accredited programs were issued ICET certificates to be awarded at graduation ceremonies. ICET is anxious for this number to be increased substantially during the coming year, especially since all applicants for certification, except ECPD graduates, will need to pass a written examination after December 1972. [This was the only report available for this meeting.]

June 1972   Texas State Tech   Lubbock
TCC Business Meeting

The Executive Committee of the Technical College Council met during the annual meeting at Annapolis to organize activities for the year. During a review of council committee activities and anticipated developments, three of the 1970-71 committees were abolished. These committees were Special Projects, Student Services, and Relations with Educational Organizations. In addition, the Relations with Industry Committee name was changed to College-Industry Relations Committee. Committees activated are listed below:

Accreditation Advisory Service Committee
College-Industry Relations Committee
Ad hoc Computer Orientation in Engineering Technology Committee
Council Historian Committee
English Speaking Union, Technical Teachers' Exchange Committee
Ethics Committee
Independent and Private Colleges Committee
Library Affairs Committee
Long-Range Planning Committee

Manpower Committee
Membership Committee
Nominating Committee
Publications Committee
Relations with Government Organizations Committee
Relations with High Schools Committee
Relations with Professional Societies Committee

The Williston Award will end with the next presentation and will not be made for 1972.

Chairman Winston Purvine reported in the absence of the council historian, E. L. Fleckenstein. There have been four boxes containing historical data that have been lost between New York and Washington. Efforts were being made to recover the boxes.

Independent and Private Colleges chairman, Walter M. Hartung, reports that in the area of higher education the extensive development of public college systems over the past several years has served to emphasize the fact that new and serious problems have been created for private institutions. The major area of concern for the independent and private colleges was determined to be a financial one. Based on several observations and conclusions, the following recommendations might be considered:

1. That no formal independent and private colleges' organization be established within ASEE at the present time.

2. That independent and private college administrators meet periodically and informally in order to discuss mutual problems and determine whatever action, if any, should be taken.

Since the National Council of Technical Schools schedules annual and, at times, semi-annual meetings at the time of ASEE and ECPD meetings, it might be suggested through the chairman of TCC that they act as host in order to provide a certain degree of continuity. It is the conclusion of the Independent and Private Colleges Committee that their *ad hoc* assignment is now complete and that the committee may be considered as having discharged its assignment.

Library Affairs Committee chairman, Frank Gourley, presented a monograph titled "Guidelines for Departmental Libraries." It included layouts, operation, and suggested materials for such libraries.

The Relations with Government Organizations Committee chairman reported on correspondence with Leon H. Blumenthal, with the U.S. Civil Service Commission: "He has answered all my questions promptly. If I do not have answers to the right questions, I believe it is because I have asked the wrong questions. The answers I have may not be satisfactory, but I believe they are accurate."

Relations with Professional Societies Committee chairman, R. E. Baird, reports that it has come rather forcefully to his attention that technical education in this country is not homogenous but is rather heterogeneous. It is pretty well understood by all what a B.S. in engineering is. But an A.E. or a B.T. in the engineering technologies means different things to different people, including the engineering technology educators. This committee was not unlike the four blind men describing the elephant. It was thought that the first thing to do preparatory to communication directly with the other professional societies was to decide among ourselves what it is that we have to tell them. A "fact sheet" with less than 20 simple facts with eye-catching cartoon illustrations has been put together. We submit this fact sheet for consideration for publication and distribution to the various professional societies.

C. L. Foster reported for the American Society of Certified Engineering Technicians. In the May 1972 issue of the *ICET Newsletter*, a questionnaire was directed to ASCET members asking for a response to the question: "In what ways have you found certification as an engineering technician to be beneficial to your career progress?" By early June, over 500 replies were received in the ICET Washington office indicating recognition through plaques, pay increases, promotions, bonuses, increased responsibility, selection for special training programs, assignment to critical projects, representing the firm at conferences, etc. Development of the new examinations is proceeding on schedule and field testing has been started. Over 30 colleges across the country have assisted in field testing the civil, electrical, and mechanical exams. ICET sincerely appreciates this cooperation and assistance in validating the examination items. New certifications are now averaging 800 per month, and the total is expected to reach 30,000 by the end of 1972. At the request of ICET, the National Society of Professional Engineers has appointed a task force to study and report in July 1972 on the impact of four-year engineering technology graduates entering the total engineering workforce. One possibility is for the BET graduates to be considered highly qualified technicians, and another is to recognize the BET degree as an additional entry into the engineering profession. Comments by TCC members on this subject are invited. You may respond to task force member C. L. Foster. This must be done immediately if you wish your opinions to be considered at the July meeting of NSPE.

1972-73 TCC Annual Report, Richard J. Ungrodt, Chair

The Technical College Council now has a membership of 68. Approximately 15 additional institutions are eligible for membership, but are currently not affiliated with ASEE nor hold membership in the Affiliate Council or Engineering College Council. The Executive Board has explored a new membership category. The Engineers' Council for Professional Development now publishes a list of those institutions which have curricula in one of the early recognition categories of "reasonable assurrance" or "candidate for accreditation." With this public disclosure, it would appear desirable to include those institutions in an appropriate membership category within the Technical College Council. The activities of TCC could be helpful to these developing institutions in moving toward full accreditation.

The Technical College Council co-sponsored the Symposium on Dual Programs in Engineering and Engineering Technology at Northern Arizona University on November 16-17, 1972. The success of that meeting produced resolutions for continuing the activities. This will aid in the articulation of the associate degree granting engineering technology institutions with others.

The Ethics Committee of ECPD submitted to ASEE a revised "Canon of Ethics and Codes of Conduct for Engineers" with no reference to the technology spectrum. The TCC chairman had previously pressed this issue both in ECPD and ASEE board meetings. The ECPD Ethics Committee does not wish to change its current report to include engineering technology, having encountered some difficulty in developing canons of ethics acceptable to the ECPD constituents. The TCC will consider the action it should take and the organizations through which it should work in order to develop possible canons of ethics and codes of conduct for engineering technologists.

June 1978 University of British Columbia Vancouver
TCC Business Meeting

Accreditation Advisory Service Committee chairman, Hoyt L. McClure, reports that for the first time there seems to be consistent referral of colleges with accreditation problems to this committee by members of the ECPD Committee and by the members of Engineering Technology Division and Council of ASEE. This change is gratifying and allows the Accreditation Advisory Service Committee to perform the function for which it was created.

TCC/ETD Long-Range Planning Committee chairman, Ernie Weidhaas, reports that the definitions of the engineering and technology profession have reached a fourth draft stage and they have been fairly widely distributed. The ultimate goal is to obtain an agreement among members of the engineering education community as well as the industrial community. Last year's report included a recommendation that contact between the TCC officers and Executive Committee chairman be increased. Although some committees are extremely active, others have never received a charge, even after repeated requests for several years. The Planning Committee recommends a "sunset" look at the present committee structure of TCC with the goal of terminating those committees which are not useful or productive. Some feel that a lean, but active, committee structure would be preferable to the extensive organization which we currently have.

Manpower Committee chairman, Jim Todd, indicates that the previous responsibility of this committee to survey all baccalaureate programs in engineering technology and publish the results in the *Engineering Education* magazine was terminated. The Engineering Manpower Commission has cooperated with us and is assuming this responsibility. Henceforth, EMC will publish this data. The main effort of the Manpower Committee, this year, has been to continue the task initiated during the 1976-77 year—establishing "state contact members." These persons have been recruited to assist in obtaining complete statistics on the various associate degree and baccalaureate degree engineering technology programs in each state, so as to provide the most accurate information available for the Engineering Manpower Commission. Although there was some improvement in the percentage of EMC questionnaires returned this year, there is still need for improvement in order to achieve our goal.

Publications Committee chairman, Frank Gourley, indicated that the Publications Committee was active this year in recruiting, reviewing, and preparing articles on engineering technology for publication in the *Engineering Education* journal (December, January, and May issues). Articles on engineering technology also appeared in the October and April issues. Engineering technology shared the spotlight with cooperative education in the May issue of *Engineering Education*. Efforts were made to provide the "Technology Education Comments" column article for each issue of the journal; however, at this time these efforts have not proved fruitful.

Relations with Professional Societies Committee chairman, John Hallman, reports that the committee finalized the listing of state and territorial PE registration requirements concerning the technician and technologist. A new survey and compilation has been initiated which is concerned with those professional societies not now a part of the ASEE. The information deals with membership of technicians and technologists and the attitude of the society toward technology education. In addition, member societies are being queried as to the status of a specific committee or individual who is responsible for technology liaison within that particular society.

Historical Research Committee chairman, Jack Spille, is convinced that to preserve our history adequately and satisfactorily it must be done by professional historians. Further, since we are concerned about a specific subject area, it will be possible to develop proposals for funding from a variety of agencies. We have taken the first steps toward accomplishing what is a substantial task and one which will be ongoing for years. Over the years we will reach our goal of preserving this important segment of American educational history.

There was a great deal of discussion relative to TCC's holding a deans' institute meeting similar to ECC. Reasons for holding such a meeting — conflicts in schedules and the development and success of the Engineering Technology Leadership Institute, which has meetings — were all discussed from the floor.

Herb Moore discussed the definitions which are going to be used in the 1980 census. The problem of not having a separate category for four-year engineering technology graduates to check was discussed at length. The job title of "technologist" was discussed and the limited use by industry of such.

January 1979 Tampa, Florida
TCC Executive Board Meeting

The TCC/ETD *ad hoc* Definitions Committee recommended that the TCC Executive Board approve "The Engineering Profession — Some Definitions," draft 6. This recommendation was unanimously approved.

More interaction between TCC and ECC is desirable. Suggestions to accomplish this included an exchange of delegates to council meetings, a special meeting of the officers of each council, and exchange of council members. Affiliation of the Engineering Technology Leadership Institute with the Engineering Deans Institute would also be helpful.

Written charges to each TCC committee should be developed. Committee members need not be TCC institutional representatives. It was recommended that the Relations with Government Committee and the Relations with Professional Societies Committee be discontinued.

June 1979 Louisiana State University Baton Rouge
TCC Business Meeting

Technical College Council chairman, Walter O. Carlson, reported that during the year enrollments have continued to grow in engineering technology education. Industry is becoming increasingly aware of the contributions engineering technology can make to their organizations. The future looks bright and the council must continue to increase its input, not only to the ASEE, but to other professional societies as well. Last year a survey was made by the Technical College Council on ECPD accreditation, and Lyman Francis and Bob Wear correlated the data. Generally, there was satisfaction among the institutions which had associate and baccalaureate degree programs evaluated during the present year by ECPD. It has been a concern of the ASEE Board of Directors that there has been only minimum coordination between the TCC and the ECC. As a first step in bringing about closer relations, Walter Thomas, Arnold Gulley, and Walter Carlson met with ECC during their spring meeting at Lake Tahoe, Nevada. We presented papers on the interface between engineering technology and engineering during the morning session. During the afternoon session, several papers were presented by ECC members on the same subject.

It is important that these interactions be continued because there is a lack of understanding which exists between engineering education and engineering technology education. Understanding can only be brought about by continual interaction among educators between the two councils. Definitions for the engineering profession, which had been prepared by an ETD/TCC *ad hoc* committee consisting of Gerry Rath, Walter Thomas, Richard Ungrodt, and Ernest Weidhaas (chairman) were considered. The definitions were unanimously approved by the board and transmitted by the TCC chairman to the ASEE Board of Directors at their meeting in February 1979. At the same meeting, the committee structure of the TCC was also discussed. This spring the ECPD Council set up an *ad hoc* committee to propose revised definitions on the engineering profession for the ECPD. An input to the deliberations was proposed definitions from the ETD/TCC *ad hoc* committee. You are all probably aware of the deliberations now in process on setting up a new organization to represent all of engineering, tentatively called the American Engineering Councils. Among the councils of the proposed AEC, the one related to accreditation would be of particular interest to those of us in engineering technology education, and we should carefully follow the developments and provide our input. During the year, 10 ASEE groups have been working on the items contained in the REETS Report. Technical College Council representatives have participated in the deliberations of the groups concerned with engineering/engineering technology interface, accreditation, the engineering society, registration and certification, and faculty development.

The Accreditation Advisory Service Committee chairman, Hoyt L. McClure, reports that the activity of this committee has increased somewhat during the year, primarily because Don Marlow of the ASEE has been referring appropriate inquiries to us. It is heartening that referral to this committee is now taking place, because that is the only way that we can carry out our stated purpose.

Publications Committee chairman, Frank Gourley, reported that the committee was instrumental in getting the monograph "Audio Visuals for Engineering Technologies" typed and printed for distribution to interested individuals. ASEE headquarters is handling requests for this publication.

June 1980   University of Massachusetts   Amherst
TCC Annual Business Meeting

The Engineering Technology Leadership Institute was recognized as having provided an important forum for ideas related to engineering technology and in the development of leadership skills. Gary Fraser moved that ETLI become affiliated with TCC. The motion was seconded. The discussion indicated that ETLI is financially self-supporting; open to administrators, faculty, and industry; members in any given year are those who attend the institute meeting; those who lead or would lead in engineering technology are especially welcome; ETLI would continue in its basic structure and direction. Sam Pritchett also spoke in favor of the motion. Motion carried by unanimous vote of those present.

TCC members were reminded of the resolution passed last year in which they objected to the proposed deletion of the adjective "engineering" from the term "engineering technology" in changes under consideration by ECPD at that time. This resolution was supported by the ASEE board and was communicated to the ECPD board. The Engineering Technology Committee of ECPD also supported the TCC position. The outgoing ECPD board did not accept the TCC position, as evidenced by the titles of the new structure which created the Accreditation Board for Engineering and Technology (ABET) and the Technology Accreditation Commission (TAC).

TCC chairman, Walt Carlson, reported that the Executive Board has recommended that the TCC by-laws be modified to provide for the vice-chairman to become chairman-elect.

In that both MET and EET department heads have expressed a need to associate within their respective disciplines, it was moved by Jim McGraw that discipline-oriented engineering technology department head groups should be formed within their respective professional societies, but not within ASEE. The discussion was supportive of such development, particularly in the EET area where the professional society is currently very active in matters pertaining to engineering technology.

January 1981 Lake Buena Vista, Florida
TCC Executive Board Meeting

Frank Gourley, chairman of the Engineering Technology Division, led a discussion on his December request to the executive director of ASEE that an auxiliary/affiliate membership category be established for the numerous educators, graduates, and employers of graduates, who support engineering and engineering technology education, but are not at present members of ASEE and may wish to have a limited involvement in activities of the society. He also suggested establishment of a system whereby engineering technology educators will be better represented in key positions and functions of ASEE. The ASEE board consensus was (1) although not initially enthusiastic, the board would consider a proposal for a partial (or single division) membership, with partial services, at a reduced fee; and (2) the board will not consider a proposal for selection of the ASEE president which guarantees selection from particular segments of the society on a regular cycle. A discussion ensued on the nomination by petition for ASEE president. Chairman Weidhaas was authorized to approach several suggested outstanding individuals to see if they were interested in running for this office.

New committees were formed and charged. The Minority Action Committee was charged to make a contribution toward increasing the present disappointingly low proportion of minorities in engineering technology. The work of this committee includes, but is not limited to, production and distribution of national radio and television public service announcements, advertisements in appropriate newspapers and magazines, the establishment of a national inventory of useful photographs, and the review of ASEE publications for improvement in this area. The Ethics Committee was charged to review and take recommendations regarding ethical considerations in the practice of engineering technology. The work of the committee includes, but is not limited to, developing a code of ethics, promoting ethical conduct and statements by the engineering technology community to reduce conflict with the engineering community and, in general, to police our own profession. The *ad hoc* Centennial Committee was charged to develop a plan for the centennial celebration of engineering and engineering technology with major effort directed to the approval by the

Citizens Stamp Advisory Committee of a U.S. commemorative stamp honoring engineering technology.

A motion was made by Jim Todd, seconded by Rolf Davey, and carried by unanimous vote to recommend changing the name of the Technical College Council (TCC) to Engineering Technology College Council (ETCC).

June 1981 University of Southern California Los Angeles TCC Business Meeting

TCC chairman, Ernie Weidhaas, reported that he has asked the TCC chairman-elect, Gary Fraser, to attend ASEE board meetings as an observer, substantially in advance of his first official attendance, to obtain maximum visibility.

A precis of the final recommendation by the Review of Engineering and Engineering Technology Studies (REETS) Committee was distributed. Three conclusions pertain to engineering technology:

1. The study showed that industrial employers have not yet evolved a consistent set of priorities for engineering technologists. The results of this survey showed the relationship of engineers, engineering technologists, and engineering technicians are and will be in a state of flux and evolution. Further studies will not solve this imperfect articulation and consequently should not be undertaken.

2. Continuing education will better prepare engineering technicians and technologists for successful careers. TCC should be charged to develop curriculum materials and methods to achieve this goal.

3. Engineering technology faculty development and prevention of obsolescence will be a continuing concern. A standing faculty development ASEE committee should be developed to continue to focus on this issue.

The council historian reported that the accumulation of the personal records and materials of Mr. Don Metz has continued during the past year. Those materials are now being cataloged by the archives staff of the University of Cincinnati Library.

February 1982 San Diego, California
ETCC Executive Board Meeting

The ETD Program of Work Committee, chaired by Frank Gourley and composed of long-range planning members from ETD, ETLI, and ETCC, met on Tuesday morning with the chief executive officers of ETD, ETLI, and ETCC, plus other interested individuals. The meeting was to discuss goals and activities of the three respective organizations.

Ethics Committee chairman, Durward Huffman, reported that over 60 draft copies of a code of ethics were distributed and only nine responses to the survey were returned. Out of this, the question was raised, "Why is ETCC preparing a code of ethics and who will enforce it?" After a lengthy discussion, the chairman charged the committee not to write a code of ethics but to observe and advise institutions, departments, and people that do not adhere to standard ethical practices in such areas as use of job title of engineers, etc.

Publications Committee chairman, Frank Gourley, reported that the May issue of *Engineering Education* will be dedicated to engineering technology. Ron Williams is open for suggested topics for the "Technology Education Comments" column and welcomes contributions on any subject related to engineering technology.

Larry Wolf reported on the possibility of creating a journal of engineering technology.

Walt Thomas led a discussion on the new HEGIS coding. A motion was made and seconded to support Walt Thomas's efforts to have the HEGIS coding modified to include a category of engineering technology programs.

June 1982 Texas A & M University College Station
ETCC Executive Board Meeting

It was announced that Mike O'Hair will assume historian duties for ETCC and ETD.

Incoming chairman Gary Fraser noted that in recent ASEE board meetings on the crisis in engineering education, it has been obvious that available data have centered on engineering. Data have not been available to document the extent of the crisis as it pertains to engineering technology, in the same manner as for engineering. Several expressed the opinion that ETCC is the most logical organization to pull appropriate data together and to provide leadership in focusing on the problems and bringing about resolution through ASEE.

June 1982   Texas A & M University   College Station
Report of the ETCC Chair

On June 25, 1981 the ASEE board unanimously approved our name change from Technical College Council to Engineering Technology College Council.

The concept of a separate journal of engineering technology (initially raised by the Engineering Technology Division in 1972) has again been suggested by Larry Wolf and others.

June 1982   Texas A & M University   College Station
ETCC Annual Business Meeting

The issue of job titles assigned by industry to graduates of engineering technology continues. Programs can lose accreditation by misrepresentation; yet even IEEE has recognized such job titles as "sales engineer, service engineer, test engineer, field engineer, and customer engineer" as typical for baccalaureate electrical, electronics, and computer technologists. ETLI and ETCC have asked ABET to recognize the jeopardy institutions face when publishing the truth about job titles assigned by industry to BET graduates and to form a committee to establish a set of typical job titles which can be used in counseling students and in promotional literature.

By consensus, it was agreed that the Ethics Committee would be dissolved and the members be thanked for their efforts.

Membership Committee chairman, Rolf Davey, reported that there were 102 full members of ETCC and 38 affiliate members as of April

1982. All potential members have been contacted in an effort to get them involved.

Minority Action Committee chairman, Vern Taylor, reported on the design of a videotape which can be used nationally on TV and radio.

Relations with Professional Societies Committee chairman, John Hallman, reported on difficulties in receiving information on membership for engineering technologists in various societies.

February 1983   Lake Buena Vista, Florida
ETCC Executive Board Meeting

Centennial Committee chairman, Ernie Weidhaas, reported that the chances of success of the engineering technology commemorative stamp appeared to be very small.

Publications Committee member Dan Hull of CORD in Waco, Texas, is compiling a monograph on resources in high technology fields.

June 1983   Rochester Institute of Technology   Rochester
ETCC Business Meeting

Publications Committee chairman, Frank Gourley, reported that the "Technology Education Comments" column edited by Ron Williams will no longer be in the *Journal of Engineering Education*. The Publications Committee has prepared a survey of publication opportunities for engineering technology educators.

October 1983   Wentworth Institute of Technology   Boston
ETCC Executive Committee Meeting

The minutes of the meeting at Rochester Institute of Technology in June 1983 were discussed. Tom Kanneman said that he is located in Tempe, not Temple. The committee voted unanimously for Tom to move to Temple, so that the minutes could be approved as read. (Just checking to see if you guys read these things.)

ASEE needs a three-year planning calendar. Both ETCC and ETLI need to provide input for future meetings.

October 1983 Wentworth Institute of Technology Boston
ETCC Business Meeting

The By-Laws Committee reported on suggested revisions. Larry Wolf, Walt Thomas, and Frank Gourley earlier chaired discussion groups relating to the revisions.

The Engineering Technology Development Committee report was presented by Ted Wisz. Much discussion followed, without agreement. Concerns were expressed that several points were contentious and that the introductory statements would bias a study done as the result of the resolution. There were several expressions that a study of the educational system for preparation of the nation's technological personnel in a changing and increasingly competitive world was needed—but for positive reasons. It was moved, seconded, and passed to commend the ETDC on its effort to date, receive the draft report, and bring additional proposals on this study to the ETCC at the CIEC meeting in Dallas in January for consideration.

January 1984 Dallas, Texas
ETCC Executive Committee

There were no corrections to the minutes of the Executive Committee meeting conducted in Boston.

The Engineering Manpower Commission report given by Stan Brodsky indicates that the technical salary survey, placement survey, and degree survey have been discontinued due to insufficient response. EMC has disbanded their Survey Committee, chaired by E. T. Kirkpatrick. This function is now provided by the Education Committee.

Minority Committee chairman, Harris Travis, will be meeting with past chairman Vern Taylor. Discussion ensued regarding minority and women's groups' disinterest in technology. It was agreed that the time was right to re-approach these groups. Stan Brodsky pointed out the unique advantage that ET associate degree programs offer persons who cannot economically commit to a longer educational program.

Ken Gowdy requested that faculty loads be studied. Stan Greenwald suggested that this should be studied in relation to tenure requirements. Chairman Fraser assigned the task to the Resources Committee.

June 1984   Salt Lake City, Utah
ETCC Executive Committee Meeting

ETCC chairman, Gary Fraser, discussed major plans for ETCC for the coming year:

A. [To build] a national resource center for statistics and information on engineering technology.

B. To investigate research in the field of engineering technology in two general areas: factors relating to educational programs and applied technology.

C. To improve the information flow between engineering technology administrators related to the accreditation process (TAC/ABET).

D. To improve communications with the Engineering Deans Council.

A discussion occurred about the possibility of ETD and ETCC using the same nominating committee so duplication of officers would be avoided in the future. It was proposed and approved that ETCC and ETD nominating committees work together to stop duplication of officers of the two groups.

June 1984 Salt Lake City, Utah
ETCC Institutional Representatives Meeting

Manpower Committee chairman, Stan Brodsky, stated that engineering technology institutions are not responding to the manpower surveys. The enrollment and degree surveys are receiving 30% response rates.

June 1984 Salt Lake City, Utah
ETCC Business Meeting

Chairman Gary Fraser stated that the ETCC by-laws had been sent to the ASEE for review. Only editorial changes had been suggested.

Larry Wolf stated that many institutions wanted to join ETCC, but due to institutional policies could not become TAC/ABET-accredited. He stated that under the proposed by-laws, unaccredited institutions could vote at council meetings but their representatives could not hold the chairman and vice-chairman positions. Several people spoke on both sides of this issue. The by-laws were amended to restrict the right to vote for institutions with at least one TAC/ABET-accredited program.

The Awards Committee had stated that there was an increase in the number of qualified people for ASEE awards. Dick Ungrodt stated that ETD and ETCC members should nominate engineering technology people for awards, besides the McGraw Award.

Ken Gowdy moved the following motion which was accepted unanimously: "We, the members of ETCC, recognize the significant organizational accomplishments during the past two years resulting from the leadership of our chairman, Gary Fraser, and we wish to formally recognize and to thank Gary for his major contributions to ETCC."

October 1984 California State Polytechnic University Pomona
ETCC Executive Board Meeting

Tony Tilmans asked to add an agenda item for the business meeting the next day. He stated that the John Wiley & Sons Publishing

Company had expressed an interest in giving an award to recognize an outstanding engineering technology program.

October 1984 California State Polytechnic University Pomona
ETCC Institutional Representative Meeting

Research Committee chairman, Ken Gowdy, stated that the committee was trying to determine the definition of ET research.

October 1984 California State Polytechnic University Pomona
ETCC Business Meeting

Tony Tilmans stated that ETD had three members at the Fellow grade — Ted Kirkpatrick, Dick Ungrodt, and Arthur Thompson. He encouraged the nomination of qualified people for this grade.

June 1985 Atlanta, Georgia
ETCC Executive Committee Meeting

Tom Kanneman reported on the progress of his committee on the closer affiliation between ETCC and ETLI. He discussed the following issues and concerns:

A. Definition of ETLI executive board membership (should not be limited to ETCC membership).

B. Chairman must be ASEE member and member of either ETCC or ETD.

C. Board members must be a member of ASEE; board must include at least one member from two-year institutions, at least one member from industry.

D. Annual ETLI meeting is essentially a non-profit proposition with institutional responsibility.

It was reported that the by-laws vote was 29 affirmative and 17 negative votes, for a total of 46 responses. In that this represents a 63% approval rate, whereas a 2/3 majority is required for modification of the by-laws, Chairman Kirkpatrick declared that the proposal to change the by-laws had failed.

Ray Sisson moved that the committee recommendation to designate Wentworth Institute of Technology as the Historical Center for Engineering Technology Education be endorsed by the ETCC Executive Committee and favorably recommended to the ETCC membership. The motion was seconded and unanimously accepted.

Gary Fraser discussed the National Resource Center for Statistics on Engineering Technology. Tony Tilmans noted that the Wentworth Institute of Technology Resource Center will have the responsibility for conduction and repository activities of research related to engineering technology (*e.g.*, Renda, Gourley, and Brodsky studies). The studies should include salary data, faculty load data, etc. Gary Fraser urged that the ETCC Resources Committee take advantage of this opportunity to identify needed information and standardization of data surveys.

Gary Fraser noted that the engineering technology faculty are falling further behind the state-of-the-art technology. He recalled the NSF institutes of the 1960's which benefitted faculty and was concerned about the mechanism for faculty development for engineering technology.

June 1985   Atlanta, Georgia
ETCC Institutional Representatives Meeting

Ted Kirkpatrick, ETCC chairman, reported that the request to move ETCC from PIC II to the College-Industry Council was deferred to the Long-Range Planning Committee of ASEE.

Ken Gowdy gave the Committee on Engineering Technology Research report. The committee is working on a bibliography of applied research projects.

Ernie Weidhaas gave the Centennial Committee report. The committee is working on a centennial stamp. The petition for the stamp will be forwarded to the Postmaster General within the next two years.

Gary Fraser suggested that this committee should take on a new charge of getting the history of engineering technology. There was discussion of preparing a document on the history of engineering technology for the centennial year 1993.

Mike O'Hair talked about the *Journal of Engineering Technology*. He asked the institutions to purchase extra copies of the *Journal* to be used for public relations purposes. He handed out a brochure describing the rate structure for bulk subscriptions.

Tony Tilmans will be sending out a letter to all institutional representatives requesting their libraries to get a subscription to the *Journal*.

Stan Brodsky gave the Engineering Manpower Committee report. He reported that he has resigned from the committee since it has moved to Washington, D.C. He stated that the commission still needed more data from engineering technology programs.

Ray Sisson gave the Long-Range Planning Committee report. The ASEE Long-Range Planning Committee has made a proposal for changing the organization of ASEE. The ETCC Long-Range Planning Committee has been a sounding-board for this reorganization.

Ted Kirkpatrick stated that the ETLI/ETCC merger question is now being considered by the officers of ETLI and ETCC. Ray Sisson stated that ETLI is presently an affiliate of ETCC. There is work being done to incorporate ETLI into the by-laws of ETCC.

Ted Kirkpatrick discussed the possibility of the development of an ETCC procedural manual. Ray Sisson stated that the manual would be developed in the next year.

Publications Committee chairman, Frank Gourley, discussed the articles that appeared in the May issue of *Engineering Education*. He stated that Ron Williams would become the coordinator of this issue during the next year, and Bill Welsh would be the editor. He asked the representatives if they preferred [to keep] the present format with the May issue devoted to articles from engineering technology or [to] spread the articles throughout the year. The majority of the representatives present preferred the present format. He stated that there is money available from ASEE for publishing monographs.

November 1985   Southern Technical Institute   Marietta
ETCC Executive Board Meeting

The draft of the report titled "Crisis in Engineering Technology" has been published in the fall edition of the *Journal of Engineering Technology*. Ted Kirkpatrick, ETCC chairman, reported that he had been informed by Mack Gilkeson that ASEE had been funded by an NSF grant to repeat the "Crisis in Engineering" survey. He stated that a similar survey on engineering technology could be included with the engineering survey. Since the engineering survey requested extensive information on research, Kirkpatrick suggested that engineering technology could best be served by separating the two surveys under one total study. This was also the consensus feeling of the board. It was suggested to request the information on M.S. degrees and not Ph.D. degrees on the engineering technology survey.

Tony Tilmans reported that the Engineering Technology Archives had been located at the Wentworth Institute of Technology. All materials, including minutes of engineering technology societies, should now be sent to the archives.

Tom Kanneman reported that junior colleges were beginning to accept high school credit for college credit. He felt that ETCC should consider the consequences of this action.

Larry Wolf wanted ETCC to consider the absence of an engineering grade for engineering technology graduates in the civil service. He also felt that ETCC should organize "SWAT" teams to talk to corporations which make decisions adverse to the field of engineering technology.

Don Stocker stated that some colleges were increasing the general studies requirements. This increase of general studies credits will increase the number of credits required for the engineering technology degree. He wanted ETCC to consider this problem.

November 1985   Southern Technical Institute   Marietta
ETCC Business Meeting

Ted Kirkpatrick, chairman of ETCC, reported that ASEE had been funded by NSF to do another "Crisis in Engineering" survey. He stated that the ETCC Executive Board recommended the use of two survey documents (one for engineering, one for engineering technology) within one large study.

Minority Action Committee chairman, Harris Travis, stated that the committee recommended to ETCC to develop a brochure for recruitment of minorities and women. He stated that more minorities and women must be recruited in the future to keep enrollments up. Dave Baker suggested that an issue of the *Journal of Engineering Technology* should be devoted to minorities and women.

February 1986   New Orleans, Louisiana
ETCC Executive Board Meeting

Ted Kirkpatrick stated that he discussed the inclusion of engineering technology on the new "Engineering Faculty and Graduate Students" survey funded by NSF, with Mack Gilkeson of ASEE headquarters. Gilkeson stated that the funds were insufficient to include engineering technology at this time. Kirkpatrick asked the ETCC Executive Board if they were willing to fund the survey with funds from ETCC. Tony Tilmans stated that the trends study in four-year engineering technology programs is performed every four years at CIEC. The above-mentioned survey could be included in this existing study. Tom Kanneman felt that the current four-year trend study was insufficient in its study of faculty. Harris Travis was concerned with the fact that engineering technology was being omitted from the ASEE survey. He made the following motion: "The ETCC chairman should make a request to the Engineering Deans Council to include engineering technology in future proposals for funds from NSF to do this survey on engineering faculty and graduate students. The Engineering Deans Council should notify the ETCC chairman in the future so the ETCC board can ensure inclusion in future surveys." The motion was seconded and unanimously accepted. Ted Kirkpatrick stated that Wentworth Institute of Technology could not do this survey this year as it had done last year. Harris Travis asked if this survey could be done by ASEE headquarters and have ETCC pay for it this year. Ray Sisson stated that he did not favor separate studies, since ASEE represents both engineering and engineering technology. It was the consensus of the Executive Board to wait one year and piggy-back onto the study next year.

Ted Kirkpatrick reported on the work of the Committee on the Education and Utilization of the Engineer. The committee has published a report called "Engineering Technology Education."

Rolf Davey, Membership Committee chairman, reported that there are now 150 full and affiliate members of ETCC. He is working with ASEE headquarters to determine the eligibility for membership of other schools.

June 1986   Cincinnati, Ohio
ETCC Executive Board Meeting

Tony Tilmans reported that the ASEE board will consider the approval for the granting of the Wiley Award for an outstanding program in engineering technology.

Steve Cheshier stated that Peter Zanetti was willing to award an outstanding laboratory award to either engineering or engineering technology programs.

Ann Montgomery Smith reported that the papers of Don Metz at the University of Cincinnati will be forwarded to the archives. Ernie Weidhaas recommended that the historian write personal letters to past leaders of engineering technology for historical materials.

June 1986   ETCC Chair's Report, Ted Kirkpatrick, Chair

The ETCC/ETLI Merger Study continues to be part of the Long-Range Planning Committee (Ray Sisson, chair) deliberations. A procedures manual for ETCC is needed. It is intended that the Gourley manual developed for ETD be adopted by the Long-Range Planning Committee.

A banking and accounting system (BASS) account has been established with ASEE headquarters in accord with the meeting vote. Our accounts are maintained by headquarters with a minimum of financial paperwork required as the officers of ETCC change over the years.

Engineering Technology Education was published by the National Academy Press, Washington, D.C. in December 1985.

November 1986 Capitol Institute of Technology Laurel ETCC Institutional Representatives Meeting

Long-Range Planning Committee chairman, Tony Tilmans, recommended a change in the by-laws to more clearly define the ETLI as a member of the ETCC. This will be referred to the ETLI for their concurrence.

A recommendation from the Long-Range Planning Committee of ASEE was presented that would change our name to ETC (deleting "college"). The ETCC recommends acceptance.

A recommendation was presented that there be two nominating committees for ETCC and ETLI. Three people would be the same on both committees — past chairs of ETD, ETCC, ETLI — plus two in either ETD or ETCC.

The SWAT teams should "Spread the Word about Technology." Emphasis should be placed on industry and corporate policies with respect to hiring engineering technology graduates.

The ASEE board voted not to endorse the World Congress planned for 1989 in Portland, Oregon. The committee suggested that possibly the ETLI could sponsor the World Congress rather than the standard ETLI, possibly in Orlando in 1989.

A study on international engineering technology education, including possibly England, Japan, and West Germany, would be beneficial to all. There is a good possibility of NSF funding.

Resources Committee chairman, Lyle McCurdy, indicated the committee is compiling a list of where faculty can obtain an M.S. degree. It is also looking into methods for faculty to gain appropriate industrial experience.

The committee recommended that we have a display at the next annual meeting for institutions to display literature for their individual programs. Also suggested was a software package information flow (what works, how it works, etc.), either through the *Journal of Engineering Technology* or the Engineering Technology Division newsletter.

Ad hoc Graduate Studies Committee chairman, Gary Fraser, indicated that the committee encourage institutions to implement graduate programs, and that any definitions of engineering technology graduate education be forwarded to Lyle McCurdy.

Minority Committee chairman, Harris Travis, reported on recommendations of the committee to include institutional efforts to recruit and attract females and minorities through ABET. Participation in minority and female programs should count toward promotion and tenure. Work should be done with publishers to eliminate discrimination.

Ray Sisson reported that the ASEE board is working IBM for funding for a software clearinghouse to receive, evaluate, review, and sell (at a nominal cost) software developed by engineering and engineering technology educators.

June 1987   Reno, Nevada
ETCC Institutional Representatives Meeting

Chairman Ray Sisson reemphasized that the McGraw Award and the Wiley Award are society-wide awards administered through ETC. The fact that we have council status, rather than division status, allows us to have a banquet on Tuesday night, which is actually an ETC banquet.

It was moved, seconded, and passed that ETC make a recommendation to the ASEE board to cut the lead time for position advertisements in the *Engineering Education News* to two or three weeks. It was moved, seconded, and passed that the chair of the ETCC Publications Committee and the past editor of the *Journal of Engineering Technology* be members of the ASEE editorial board.

Membership chairman, Rolf Davey, indicated that at present there are 102 paid-up full members and 37 affiliated members in the council.

Tony Tilmans discussed a program whereby new faculty members in their first five years of teaching could receive two years of free membership in ASEE, one half to be paid by ASEE, the other half to be paid by the member institution. At the Engineering Deans Council meeting this year, 60 institutions signed up for this program. He recommended that we should consider the same, publish the information, and make up a membership brochure which would include a listing of participating institutions. He also suggested the possibility of waiving ETD membership fees during this time.

Ray Sisson informed us that ASEE board approval of increasing individual membership fees to $50 has to be approved by the membership.

Ray also informed us that the board has approved an increase in institutional memberships from $435 to $550. This does not have to be approved by the membership and will be in effect next year.

Accreditation Relations Committee chairman, Earl Gottsman, reported that ABET is discussing new criteria for faculty in engineering technology. The committee is concerned with these criteria. The proposed criteria are that all faculty must have at least a B.S. in engineering or in engineering technology and for two-year programs the majority must have an M.S. For four-year programs, 75% must have an M.S.

February 1988 San Diego, California
ETC Board Meeting

Ray Sisson, ETC chair, declared that the proposal for ETC to join EDC in the Federal Liaison Office was approved.

Ray Sisson distributed a copy of the proposed brochure on engineering technology in ASEE. He asked that all comments and the ballot be returned to Frank Gourley by March 15.

Earl Gottsman, Manpower Committee chairman, reported the EMC is looking at different surveys, attempting to refine data and produce more consistent results.

Resource Committee chairman, Don Stocker, has a list of 40 graduate schools for engineering technology graduates. He will have the report ready for the June meeting.

Minority Action Committee chairman, Harris Travis, announced that the committee has managed to schedule a plenary session for the June meeting.

Chairman Ray Sisson reported that the president-elect of ASEE is looking for nominees to all committees. Ray especially solicited nominations of ETC and ETD members to these committees.

October 1988 Purdue University West Lafayette
ETC/ETLI Business Meeting

Tony Tilmans, ETC chair, met with Dale Parnell, AACJC, and they are extremely interested in working with us to work on a joint project; *i.e.*, student pipeline, minority effort. Tony visited the Department of Education and opened the door for further work. We should look at student and faculty pipeline issues. We should form our own ETC task force. We should be working on assessment and accountability. We need to continue to study faculty data, resource data, trends, etc.

Federal Liaison Committee chairman, Bill Troxler, indicated that NSF was interested in funding a workshop of industry and education leaders to define the role of the BET in industry.

January 1989   New Orleans, Louisiana
ETC Executive Committee Meeting

The Federal Liaison Committee chairman, Tony Tilmans, reported that Ann Lee Speicher, ASEE, is doing an excellent job for ETC. She was able to request added language to the Tech Prep bill to clarify engineering technology participation.

Tony Tilmans reported on the NSF two-year college workshop. Engineering technology will be included in all future requests for proposals.

Engineering Spectrum Committee chairman, Ray Sisson, distributed copies of the committee's draft statement. Hal Roach expressed concern that the draft may not reflect the consensus of ETC. After much discussion, it was moved and seconded that the ETC board should not forward the draft statement to the ASEE board. The motion was passed unanimously.

The ASEE editorial board has made a final decision not to dedicate one issue of *Engineering Education* to engineering technology.

Membership Committee chairman, Rolf Davey, reported that contacts are being made with all community and junior colleges (not having TAC/ABET programs) informing them about ASEE and affiliate membership in ETC.

Graduate Studies Committee chairman, Al McHenry, reported that he would like this committee to be a clearinghouse for all graduate programs relating to engineering technology. He will be sending out questionnaires this spring so that his committee can establish a catalog of programs.

Steve Cheshier reported that his sub-committee had completed its research and recommends that "old gold" be designated as the certified color, nationally, for engineering technology. If the degree is actually a Master of Science, it would be permissible to use regular gold, since that is the generic color. If a college uses engineering orange, that would also be acceptable.

June 1989   Lincoln, Nebraska
ETC Business Meeting

Bill Troxler and Tony Tilmans reported on pending legislation that would change the Carl Perkins Act terminology from "vocational-technical" to "applied technology." Bill HR7, as passed, would provide $200 million for the Tech Prep initiative.

Tony Tilmans reported that Karl Willenbrock feels confident that NSF will fund the Engineering Technology/Engineering Interface Workshop that has been proposed. There was considerable discussion related to the position that the ET community might take at the workshop. Larry Wolf reported that ABET has appointed a committee to study the Engineering/Engineering Technology Interface. Ray Sisson reported that the Spectrum Committee is still studying the issue and has not issued any official report. Dave Baker reported that the New York State Licensing Board stopped action on ET issues until ASEE finished the study. Ray Neathery made a motion to have ETC send ABET a letter emphasizing the fact that the report they have is strictly a draft and not a final document. Rolf Davey made a motion to have ETC go on record as supporting the concept of the workshop. Don Gentry proposed an amendment to the motion that would add: "The existence of engineering technology should not be an agenda item." Ray Neathery proposed a second amendment to the motion to include the statement: "The slate of ET members nominated for the workshop must be approved by ETC members." The main motion was then approved as amended.

Publications Committee chairman, Frank Gourley, gave a detailed report on the many activities of this committee (copy on file with the secretary).

Membership Committee chairman, Rolf Davey, distributed a report showing that as of June 1989, 130 institutions were members of ETC (101 full members, 29 affiliate members). The committee is actively soliciting new members.

The Relations with Professional Societies Committee chairman, Elliot Eisenberg, indicated that the Electrical/Electronic Engineering Technology Department Heads Association is exploring the formation of an electrical engineering technology society within IEEE. The committee strongly encourages the engineering technology community to actively pursue appointed or elected positions within their respective professional societies.

Accreditation Relations Committee chairman, Earl Gottsman, distributed an annual report outlining the many activities of this committee.

Paul Rainey made a motion to have ETC approve a list of reviewers for technical articles for proceedings that are submitted by the department head groups. The Publications Committee was charged with developing a list of reviewers for non-technical articles.

Steve Cheshier reported on the status of an ET discipline hood color for academic regalia. Unless a discipline is entirely new and distinctive, no new color will be designated. For a field not included in the code, the committee encourages institutions to use the color of that listed discipline most closely related to the field. For engineering technology, engineering (orange) would be the obvious choice.

November 1989 Norfolk, Virginia
ETC Business Meeting

Bill Troxler reported regarding the federal liaison activities. The Tech Prep bill has been passed by the House and is out of committee in the Senate and will probably go to conference early in 1990. ETC representatives are encouraged to write their senators on the issues outlined and thank them also for including engineering technology in the bill.

John Wiley & Sons will no longer publish ET textbooks and is therefore withdrawing its support for the Wiley Award.

The Long-Range Planning Committee chairman, Steve Cheshier, recommended that the Resource Committee take responsibility for continuing the longitudinal study that was started back in the mid-70's. ETC should make a greater effort to promote affiliate membership and to then encourage accreditation to qualify for full member status. The Wiley Award needs to be restructured. It is important to keep the award open to avoid having to go through the approval process again at the ASEE board level. Several possibilities were discussed for funding the award. Having it endowed would be preferred.

Publications Committee chairman, Frank Gourley, indicated that four surveys are currently in process: (1) textbooks, (2) AV materials, (3) lab experiences, and (4) ET institutions and programs. Also efforts are being made to solicit articles for *Engineering Education*, other society publications, and popular business magazines.

June 1990   Toronto, Ontario, Canada
ETC Business Meeting

Bill Troxler reported for the Membership Policy Committee. The goal is to increase membership. When the dues were raised, industrial memberships lowered. Services to members need to be increased to increase individual memberships. The following recommendations are made: (1) a sliding scale for dues; (2) two at-large corporate members of the executive board should be appointed by the president of ASEE; (3) new industrial members should be recruited through advisory boards; (4) recruiting faculty and graduate fellowships should be encouraged at meetings; (5) sessions should be given to assist faculty to renew technical expertise, in proposal writing, and with tenure track consultation; (6) federal briefing sessions should be held to inform members of funding opportunities, etc., like the NSF one this year; (7) ties need to be strengthened with other educational groups; (8) ASEE publications are confusing and misleading and need to be redone. Twelve percent of ASEE members are retired, honorary, or life members with reduced rates, and this number is expected to grow. There needs to be a tie between active and retired member dues. Retired should be changed to senior membership. Life membership should be based on age and years of membership plus service, to use talents still.

Curt Tompkins, president-elect, welcomed the group and spoke of his eagerness and openness to learn more about engineering technology. He would like to see vision and mission statements for each group in ASEE. He has set as a high priority improvement in the quality of member services by ASEE.

The Frederick J. Berger Award was approved. Committee selection — nominees must be over 50% teaching and not from the same school as the current/past ETC chair.

Publications Committee chairman, Frank Gourley, indicated that the *Directory of Engineering Technology Institutions and Programs* is available from ASEE headquarters. Surveys of textbooks, of lab experiences, and of AV materials are being done and will be used to create databases.

Longitudinal studies have been done every four years, three times. The Resource Committee revised questions, and the studies will be sent out this summer. A report will be made at CIEC.

The ASEE *Institutional Program Directory* currently lists four-year programs only which pay for the entry. In the future, two-year programs (associate degrees) may be included.

November 1990   Kansas Technical Institute   Salina
ETC Business Meeting

Long-Range Planning Committee chairman, Dave Baker, indicates that ETC needs to develop mission and vision statements. Tony Tilmans indicates that ASEE is interested in including ET programs in the *Research and Graduate Study Directory* and in the *Instructional Programs Directory*. The associate degree programs are to be included in the future.

Membership Committee chairman, Rolf Davey, reports that a new computer system at ASEE headquarters will make the Membership Committee's job easier.

June 1991   New Orleans, Louisiana
ETC Executive Board Meeting

It was noted that the engineering technology educator community contributes about $150,000 to the society each year and gets back about $3,000 (ETC and ETD). This year $150 minimum will be allocated.

June 1991   New Orleans, Louisiana
ETC Business Meeting

Relations with Professional Societies Committee chairman, Elliot Eisenberg, indicates that some progress has been made with ASCE for recognition of engineering technology. In ASME, an award is to be given. The technology community needs to more actively participate in professional societies.

Representatives were asked to check if their schools are subscribing to the *Journal of Engineering Technology*.

Publications Committee chairman, Frank Gourley, indicated that the bibliography of engineering technology education articles is being compiled by Marilyn Dyrud. Three other surveys are in process — surveys of textbooks, audio visual materials, and laboratory experiences. The *Directory of Engineering Technology Institutions and Programs* is available from ASEE headquarters.

The Committee on Society Awards (Wayne Hager, chair) is a new committee whose charge is to encourage nominees and to promote recognition of active individuals, keeping track of who should be recognized.

Joe DiGregorio will chair the new International Committee. The committee will begin by surveying ETC and ETD for current activities.

Long-Range Planning Committee chairman, David Baker, indicates that a mission/vision statement has been drafted. The ETC By-Laws/Procedure Manual needs revision.

ASEE director, Frank Huband, indicates that the *Prism* preview issue is out. Publication has been changed to desktop with a cost reduction from $44 per page to $9 per page. A new student magazine is being developed with ads to break even. It will be distributed to students from the tenth grade to the master's level. This will increase visibility and reach.

January 1992   Las Vegas, Nevada
ETC Executive Committee Meeting

Steve Cheshier, chair of ETC, reported that, since the actual membership in ASEE is about half of what it was thought (5,000 instead of 10,000), membership fees will have to be increased to increase revenues. In June 1992, there will be six ET people on the ASEE board. In a recent survey of the ETC institutional reps, 24 indicated they would like to see only one ETC business meeting per year (at the annual ASEE meeting), while 20 preferred two business meetings per year; if a second meeting is held each year, 19 said they would prefer it at CIEC, while 15 said they would prefer it at ETLI. Cheshier indicated that these results were inconclusive. He suggested that the ETC board ask the ETLI board to consider moving ETLI to the front of CIEC and reducing the ETLI from two and one-half to one and one-half days to cut travel and other expenses for ET people who now must attend both meetings. After considerable discussion of the pros and cons of this suggestion, Cheshier moved and Emshousen seconded the motion to make these two suggestions to the ETLI board. Mike O'Hair suggested a "friendly amendment" to discontinue the ETC board meeting at ETLI, but to still have time at ETLI set aside for ETC committee meetings. Fred Emshousen amended the motion to suggest that the 1992 ETLI at OIT be held as planned, and that the movement to the front of CIEC begin the following year, in deference to all the work and planning that had already gone into this year's ETLI at OIT.

Earl Gottsman reported that after receiving several "death threats" from a group of militant feminists led by Jane Downey, the "Engineering Manpower Commission" has decided to change its name to the "Engineering Workforce Commission."

January 1993   Lake Buena Vista, Florida
ETC Executive Board Meeting

ETLI has proposed that a separate meeting in the fall not be held this year. Rather they will conduct themed workshops on the front-end of the CIEC conference next year. They will need to be sponsored by ETD. Since ETLI is a part of ETC, the ETLI board asked the ETC Executive Committee to endorse this plan.

Mike O'Hair gave a report on the ETC/ETD centennial plans and indicated that the history would be about a year late.

Chapter 6

Oral Histories

Introduction
by
Alexander W. Avtgis
Wentworth Institute of Technology

No history of engineering technology education would be complete without the reminiscences of those early leaders who made significant contributions to its emergence. The considerable effort of those who have researched and written the formal history has produced a document that weaves together the threads of activities in different places and at different times. But the narratives presented in this chapter convey, in very personal and real terms, the conflicts, frustrations, and accomplishments which brought engineering technology education from trade-school status to a recognized professional college-level discipline and one of the largest divisions within ASEE.

The Centennial Committee generated an exhaustive list of ET leaders going back to the early years. Of these, eight were located and have been interviewed for this oral history project. Their memories

 See Appendix I for photo captions and credits.

serve to remind us of the many others who worked with them and made significant contributions as well.

We are deeply indebted to those individuals who volunteered to conduct the oral interviews. Their efforts in scripting the interviews, asking follow-up questions, and producing the written documentation were essential to the success of this project.

The Centennial Committee invites you to salute in gratitude these eight engineering technology leaders who share their personal experiences with us:

> Walter M. Hartung
> Lawrence V. Johnson
> Michael C. Mazzola
> Hugh E. McCallick
> George W. McNelly
> Winston D. Purvine
> Richard J. Ungrodt
> Eric A. Walker

Editor's Note: The oral histories have been edited for consistency and to delete repetition. To improve readability, occasionally words have been added; these are indicated by brackets in the text. The original audio tapes (and L. V. Johnson's videotape) and verbatim transcriptions of the histories are located at the Engineering Technology Archives, Wentworth Institute of Technology, Boston, Massachusetts.

On occasion, the interviewer may mention the "questions on the list." This refers to a list of five basic questions asked of all interviewees:

- *How and when were you involved with engineering technology in ASEE?*
- *What, in your opinion, were the most significant issues and activities in engineering technology that ASEE dealt with?*
- *What progress do you feel that engineering technology has made?*
- *What in engineering technology's development did you not anticipate to occur over the years?*
- *In your view, where is engineering technology today compared to the 50's and 60's?*

Of course, interviewers naturally asked follow-up questions, and interviewees did not confine their comments to these issues.

Walter M. Hartung

Former President Academy of Aeronautics

Interviewed by
Robert J. Wear
Former Dean
Academy of Aeronautics

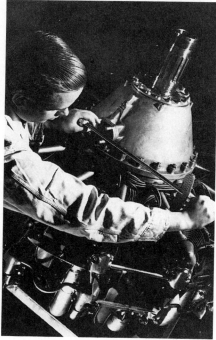

Walter Hartung received two degrees in aeronautical engineering and an M.A. and a Ph.D. in education from New York University. In 1928, he began his career as a detail designer of light aircraft. He was one of three designers responsible for the Granville Racers flown by James A. Doolittle and others.

After five years in industry, Dr. Hartung began teaching at the Casey Jones School of Aeronautics, where he was responsible for developing engineering, design, and special war training programs for Army Air Corps technicians during World War II. After the war, he was executive vice-president and dean at the Academy of Aeronautics, Casey Jones' successor.

Dr. Hartung has been active in many engineering

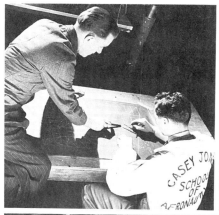

professional organizations and in 1961 chaired the U.S. Technical Delegation to the Soviet Union.

Robert Wear: This is for ASEE, Dr. Hartung. They selected four or five people to talk and make a tape of what went on in the early days, and you were one of them.

Walter Hartung: I am very pleased to make this interview for the early days of ASEE, and this goes quite far back. So I have to scratch my memory to get to it. But actually ASEE took in the technical institute people in the early days and didn't quite know what to do with them. We had a committee of 21; I guess it was named for that (Committee of 21) because there were 21 members on it. They were a sort of stepchild of the ASEE, and they gradually took over the accreditation of their programs and things like that.

There were some very interesting people in charge of that time; one of those was Karl Werwath, who was a moving light in the development of technical institute approval and accreditation. That went a long ways [to] getting us into the position where we were. Now the problem was that the air technical institute schools had a title, which, in most cases, was pretty close to engineering, and that was one of the biggest rubs in the business. The engineering schools did not particularly want their name to be used as part of the school name, and I can see why they wanted that, but, nevertheless, we had to develop too.

Now at the Academy of Aeronautics back then and even the Casey Jones School of Aeronautics, we developed some pretty good engineering-type programs. But engineering technology was a name given to them and perhaps that should have been a name that was given to the engineering program, but it was changed quite a bit. The first accreditation was developed as an institute of technology program, or technical institute-type program, and the engineering accreditation body had a sub-committee which checked these schools and developed an accreditation procedure.

Actually, the accreditation [was] pretty difficult as far as these schools were concerned. They gradually developed pretty good programs, and they got accreditation and gradually developed into the Technical Institute Division. That was an outgrowth of the Committee of 21, and then it became the Technical Institute Council. Then the group was recognized a little bit more.

The National Council of Technical Schools was an interesting program. That was a council of all of these schools out of which the Committee of 21 operated. National Council of Technical Schools, in lieu of no other means of accreditation, developed the National Council of Technical Schools accreditation proceedings: DeVry, Central, Academy of Aeronautics, Wentworth, the Rietzke school, and Capitol Tech.

They developed quite an accreditation program which [meant] they were eligible for veterans' funds out of World War II. Without that, they could not have operated. The National Council of Technical Schools was probably one of the first accrediting agencies in the technical institute business. They developed agencies which, of course, eventually led to the Technical Institute Council, which had pretty good status as far as ASEE is concerned. That came about in the mid-1960's.

There was a lot of tension in the development of recognition for the technical institute, and the engineering deans cooperated, but not to any great extent, until later on, when they had to give the programs themselves. Accreditation was the major problem. In other words, at the end of World War II there were lots of veterans going back into training. They selected the programs which were paid for by the government. But the government would only pay to accredited programs. That word "accreditation" was a real stickler, and it was right that they had it so, as a measurement of whether they paid for the tuition or whether they didn't. And without that, these schools couldn't operate. So the National Council developed an accreditation agency which was recognized by the government, and we went through many, many years before we became accredited through ASEE.

That was, of course, under ASEE; the Technical Institute Council was then established. The people who were selected to do the accrediting essentially came out of the National Council. Most of them did; it was a carry-over. The National Council people just kept doing what they were doing, and then they operated under the Engineers' Council for Professional Development. When ECPD recognized these schools, we were in pretty good business.

Wear: **Who were all the people who were involved?**

Hartung: As I mentioned before, Karl Werwath, Curly Foster, Rietzke, Thompson from DeVry, Joe Gershon (he came later), and then there was Hugh McCallick in Houston. He went to Capitol for a year or so, and then he went back to Houston, but he was originally with

Houston. Russ Beatty from Wentworth, of course. I was associated right from the beginning. Casey Jones was involved at this point. I worked with Casey on this, and Casey was involved as head of the institute for some time. I did most of the work for him. Casey Jones was very much involved in this.

In ASEE, the division came in before the council. That's right. The National Council was kind of an off-shoot. The National Council was taken into ASEE. I guess it closed its doors when we were out in that meeting in San Francisco.

Wear: **What closed its doors?**

Hartung: The National Council. There was no need for the council anymore. Of course, then we saw the input from the public schools. They were starting to grow at this time. They were starting to put in two-year programs. Some of those were the New York State schools; Penn State was a very big one; Purdue; DeVry, from the beginning; Morgan Institute of Connecticut; Wentworth; and the Academy of Aeronautics (Casey Jones School had closed its doors in Newark).

It was after the war, of course, when we all came back and developed these new programs for the veterans. Veterans were the only ones [who] could go to school at that time. The whole emphasis of recognition of any schools at that time was through the payment of tuition by the government, and it had to be paid to an accredited institution. The accrediting institute for our schools was the Engineers' Council for Professional Development, which is now the Accreditation Board for Engineering and Technology. There was just no other way around it but to get an accreditation.

Of course, ASEE is not an accrediting body, as you know, but it was the only group willing to do any of this at that time. The other professional societies were not involved in engineering. But they really didn't send any people for accrediting technical institution programs. Among the early college people was Dean Hammond of Penn State, who was a big mover in this group. I wouldn't want to leave his name out. He was a big mover for getting recognition of technical institutes of the two-year colleges. Then along came Ken Holderman; he was to Dean Hammond as I was to Casey Jones. That was the best way I can explain it, but Dean Hammond was a great believer in the technical institute-type program. Of course, Penn State had a series of 18 or 19 schools. They have now expanded to include four-year programs.

Wear: When did the professional societies get into the act?

Hartung: ECPD? That would that have been in the mid-60's. It seems to me that you and I were both on ECPD during the mid-60's period. We were then starting to conceive the first of the professional societies such as IEEE and the civil ASCE and the chemical people. The chemical people were very much four-year oriented, and they did not [look] too kindly on the accreditation of two-year programs because they didn't figure [they were] strong enough to warrant accreditation. Of course, it was cutting into their field.

We have had our go-around with the professional engineering societies. Karl Werwath worked on that for quite some time to get industrial recognition for the two-year program graduate. I don't know if Karl really accomplished that, but what they actually did was the graduates kept getting experience, and then got their professional license through experience rather than college.

You see, a graduate of an engineering college can automatically go for a professional engineer's license, but it wasn't true of two-year program [graduates]. Two-year programs plus a couple of years out in industry and then they would qualify and take a test and get a professional license. Of course, then the professional societies really came in and that's when ECPD stopped and ABET picked up.

Wear: Well, ABET was more of a change of name than anything else, wasn't it?

Hartung: Well, yes. But then we had the expansion of the societies. There were five original professional societies, and then they expanded, and I don't remember what the first five were. But that, of course, would be out of ECPD. And then ABET came along, and that's when there was that big fight. You remember, over what it was going to be called. The "Accreditation Board for Engineering and Technology" was the final outcome, which put us on even footing. But that didn't occur until a few years ago, perhaps in the late 70's; it must have been after that.

A lot of this society activity was recognized, I guess, by the McGraw Award, because most of these people [who] were involved in this early activity were later recognized and received the McGraw Award. I don't think it had any relationship to their association with the various activities, except that they were the active people. The McGraw Award gave the active people credit for the movement, so to speak, in the technical institute movement.

Lawrence V. Johnson

Founding Director Southern Technical Institute

Interviewed by
William D. Rezak, Dean
School of Technology
Southern College of Technology

After receiving a B.S. in both engineering and physics (1930) and his M.S. in physics (1931) from Ohio State University, Mr. Johnson quickly entered the field of technical academics. His career began at Georgia Tech in 1942, where Mr. Johnson progressed from physics instructor to director of its Daniel Guggenheim School of Aeronautics. By 1948, Mr. Johnson, as director of Georgia Tech's Technical Institute Program, was laying the foundation for Southern Technical Institute.

At its inception, Southern Technical Institute enrolled 117 students, which grew to well over 900 students during Larry Johnson's tenure. Today, the college has some 4,000 students.

Although he retired in 1963, his leadership and the foundation laid by Mr. Johnson and his

team created the groundwork for what is now the Southern College of Technology.

William Rezak: We're here today with Mr. Larry Johnson, the first director of the technical institute, Southern Tech's predecessor. Mr. Johnson began his career as an instructor of physics at Georgia Tech. He was appointed assistant professor of aeronautics in 1942 and served as acting director of aeronautics from 1944 to 1947. Mr. Johnson took a leave of absence from Georgia Tech in 1945 and 1946 to teach electrical engineering at American Biarritz University in Brunswick, France. During World War II, he coordinated the Georgia Tech Civil Aeronautics Training Program. He became director of the technical institute on October 15, 1947. His leadership led to the establishment of the Southern Technical Institute. At that time, Southern Tech was part of Georgia Tech's Engineering Extension Division.

Larry Johnson assumed a major role in explaining the purpose of technical institutes to educators, industry, and parents. He sponsored and helped prepare the first motion picture on technical institutes, entitled "The Technician in Industry." This film, produced by Southern Tech's faculty, was shown to high school students and civic groups all over the Southeast and in technical institutes around the U.S. The Ford Foundation used the film in its development of the technical institute programs in foreign countries.

Mr. Johnson was a member of and also chaired the Engineering Technology Accreditation Program of the Engineers' Council for Professional Development. He won the coveted James H. McGraw Award in Technical Education in 1963.

Mr. Johnson, it's a pleasure to be with you here today. I'd like to ask you about yourself, the development of Southern Tech, and engineering technology education in general. Please give us a little insight into your own education and professional experience leading up to your involvement with the college.

Larry Johnson: As you said, you have covered my background in education. I was in the Guggenheim School of Aeronautics when the interest came for the technical institute program. At that time, Blake Vanleer, the new president of Georgia Tech, talked to the Associated

Industries of Georgia and asked them what he could do to serve their organization. They came back with the statement, "You are educating the officers of engineering, and we can educate the privates of engineering. What we need is someone to train the sergeants of engineering technology." So he came back with the idea of setting it up; that was in 1945. He created a committee to study the program, and in 1949 we had the answers for three items.

One was the location. At that time, we had to study three locations, but it turned up at the last minute that the Naval Air Station in Atlanta was about to be vacated. We were able to get seven buildings and a dormitory at the magnificent rent of one dollar a year, plus maintenance. We never did much maintenance.

The next job was to sell the idea to the Georgia Tech faculty, who opposed it very much. They did not want it because they thought that it was a vocational program. We also had a hard time selling it to the Board of Regents because they also thought it was a vocational program. However, it turned out that Vanleer decided to put the program in the Georgia Tech night school. And, because the faculty had no control over it, that was done.

In selling the Board of Regents, who were hard up at that time, he made a statement the technical institute would be self-supporting in two years. This was a total calamity for us because, after two years, we were far from able to support ourselves. In fact, Georgia Tech did give us $65,000 for the first year and $75,000 for the second year. But the third year, the G.I. (Georgia Institute) program began to terminate and they did not have the money. So the Board of Regents finally considered us very carefully and decided to give us some support.

At that time, the first support was $64,000 and by 1952 increased to $509,000, so we did very well on that. My first job was to go out and establish the program at the Naval Air Station, which consisted mostly of dormitories and one dining hall with a place to sleep. Anyhow, we went out and began work on January 2, 1948: five of us to create the laboratory, the catalogue, the program all the way through, and to begin hiring the regular faculty.

Rezak: **Mr. Johnson, please give us a little background on your professional experience, education, and background prior to coming to the technical institute in 1947.**

Johnson: I graduated from college in 1931. I took that job in 1947. So there was a long time between. In between, I graduated from Ohio State. I began as an instructor in physics at Georgia Tech.

Rezak: So you had a degree in physics from Ohio State?

Johnson: I had a degree in physics and a master's in physics from Ohio State.

Rezak: So, you were teaching in physics and aeronautics at Georgia Tech through the 30's and 40's.

Johnson: Yes, from the 30's to the 40's — about 15 years.

Rezak: I see. All right, now you mentioned that you started in the first part of 1948, I guess, with the fine folk developing the institute. Who were the people involved with you at that time?

Johnson: There was Dean Mattlicks, Dean Carroll, Bob Hayes. Bob Hayes just retired out there, I think.

Rezak: Bob Hayes has been retired for three to four years, I'd say.

Johnson: But he was one of our first employees. I have forgotten the rest of them. After 21 years, I forget some of those things.

Rezak: Why was the institute created in the first place? What was the driving force behind that vision?

Johnson: The reason that Southern Tech was developed was at the request of the Associated Industries of Georgia. In 1945, the new president of Georgia Tech, Dr. Vanleer, addressed the AIG and asked them what he could do for them. He came back and appointed a committee, of which I was a member, to study the program and make recommendations.

As I said before, I believe there was much opposition to the establishment of the program from the Georgia Tech faculty and the Board of Regents. So he had to overcome that. He overcame the faculty [resistance] by putting it in the Engineering Extension Division. He told the Board of Regents that he had enough money in his surplus account to operate it for two years and that it would be self-supporting after two years, which certainly was not true.

Rezak: **Where did you get your students and where were they recruited?**

Johnson: Among other things, we went around to all the high schools in Atlanta. Our first students were 110 veterans. At that time, the veterans were supported by the government and were looking for places for good education. We had 10 civilian students, which is certainly not very many, and one girl, which totals about 117 or 118 students.

Rezak: **So there were even women in the program that early on?**

Johnson: Yes, and that woman was a very good-looking young lady. She was the first woman to go to Georgia Tech; she got a lot of publicity because they came out with a headline of "Tough Tech Female." She was interviewed by many of our organizations, and she was written up in the *Atlanta Journal* magazine section and things like that. She produced about $10,000 worth of advertising for us.

Rezak: **That's great. I'm sure that was a big help in recruiting more students?**

Johnson: We had at the end of the second year about 400 students.

Rezak: **Mr. Johnson, what kinds of jobs did Southern Tech graduates earn after they graduated in the early years?**

Johnson: A lot of them went into the Georgia Highway Department, which was hard up for technical people, and a lot went into Atlantic Steel Company. In fact, one became president of Atlantic Steel after three years. Others went into the telephone company, highway department, and other places. We had no trouble obtaining jobs. They started out with ridiculous salaries for now, but they started out at about $3,000 to $3,600 a year.

Rezak: Now I know that Tau Alpha Pi Honor Society nationally was started at Southern Tech. Can you tell me when that occurred? What brought that about, and how did it evolve?

Johnson: We had an instructor there named Jesse Defore, who was very much interested in the quality of the programs and honor students. I think he started the program in about (I'm guessing) 1957.

Rezak: What were the educational programs the institute offered in its early years?

Johnson: Well, unfortunately, we started with too many. We started with, I think, 12, and we finally reduced to seven. I think electrical, mechanical, industrial, civil — I don't remember the rest of them.

Rezak: How did the college come to move from its location at what is now the Peachtree DeKalb Airport to the campus in Marietta?

Johnson: Well, we started there in '49 and stayed there through '62 or '63. At that time, they wanted us to get out of there so they could use the old Naval Air Station as the DeKalb Airport. And so we were encouraged to move, although we were not forced to move. They associated us in Georgia, which got us started in the first place. President Vanleer went to the governor, and he got us $3.5 million [for] new buildings and supplies. And he got us land, 120 acres, I believe, out in Marietta, which used to be the old Marietta Poor Farm. We began building in 1961, and I think we moved out there in '63.

Rezak: What was the role that Southern Tech played nationally in the evolution of engineering technology education?

Johnson: The big thing that we did was that most of our faculty [were] from engineering education in the first place. So we inadvertently (or not) put in a lot more technology than most of them. After we got started, Lockheed out here was very positive and advertised for us. We put out a person who was really a semi-engineer. Between that and the moving picture we wrote that was shown around, we really programmed the technical institute and Southern Tech all over the country.

Rezak: And were you all who were formulating the institute in the earlier years very active in American Society for Engineering Education at that time? Tell us about that and how the Engineering Technology Division of ASEE was formed and grouped.

Johnson: Well, I've been a member of ASEE since the second year I went to Georgia Tech, in 1932. I was in the first Committee of 21 for the technical institute programs. We had a hard time with ASEE, just like we did with the Georgia Tech faculty, because they were an engineering group and were hard to convince that we were anything but a vocational program. So it took a lot of energy, determination, and talk to convince ASEE to put the accreditation of technical institute programs in their program.

My only reaction to that is that I think you will get more and more of the master's degree programs, in five years, maybe six. I don't know if you'll go to a Ph.D. program or whether you'll need it or not, but it is just like any other program: as time goes on, I guess you'll need more technical, more management theory, more English, and everything else. Eventually, education will be the salvation of America. I think technical institutes will be the salvation of a lot of engineering and technical work.

Rezak: How about some humorous events that occurred early on when the campus was first started and the programs were first being formed. Any thoughts here?

Johnson: One of the funniest things we did was put on "H.M.S. Pinafore," which was quite popular. We showed it two or three times around the area. But I don't know of any humorous things in particular.

Rezak: So there was a full complement, perhaps, of student activities such as theater and what-not for people to relax and refresh themselves outside of the classroom and laboratories?

Johnson: Yes. Of course, they had a basketball team from the beginning.

Rezak: They had a basketball team from the beginning? Is that right? And who did they play back then?

Johnson: We played a lot of high schools and we played some colleges. We played a college up in Chattanooga and some places like that, but we were not in the big league. We had a man who was real good at it, Frank Johnson. He left us and went to Lockheed to set up their program for a long while and finally died. He was a big heavy fellow. We never did play football. I think that was too expensive for us. Unless you've gotten into it since I left.

Rezak: No, no. Same story. **What other recollections do you have that you might like to share with us about the development of engineering technology education in the American Society for Engineering Education or about the development of Southern Tech?**

Johnson: Well, as I have said previously, the biggest development, I think, in ASEE and accreditation is the recognition of the technical program, first in the two-year degree, then in the four-year degree. I hope they'll soon recognize it in the master's. I think [this] benefits all technical institutes.

Moving on about 10 more or 12 more years, we had another program we spent a lot of time establishing the accreditation for, the four-year technical institute program. In fact, I was on the board of ASEE when we put that through. Frankly, I think the next program we'll have in ASEE with accreditation will be a master's degree.

Rezak: **When did Southern Tech begin to offer four-year baccalaureate degree programs in addition to the successful two-year associate degree programs, and how did that occur?**

Johnson: The pressing issue [was] that they needed more training. We couldn't get all they needed in two years. And after we got the two-year program going, the students themselves felt that they needed more education. They needed the baccalaureate degree and the recognition for their work more than the associate degree.

Rezak: **Could you tell us a little bit about how you think Southern Tech has changed over the decades while you were involved with it, and then afterwards, please?**

Johnson: Well, unfortunately, I have been out of the whole program for the last 22 years, and I would not be very well qualified for that.

I think there will be more and more technical institutes all of the time, but I don't think that these high school programs we have established all over the state and called technical institutes will ever amount to much more than vocational programs. I know we changed the name of Southern Tech to Southern College of Technology because of that.

Rezak: **One of the troubling aspects to me, as a current professional in engineering technical education, is the challenge of getting industry to recognize, in some cases, the job classification of the engineering technologist. One of the challenges graduates experience is going to work in engineering positions in competition with graduates of engineering programs from other institutions. Any thoughts that you can share with us on the industry approach to engineering technology education?**

Johnson: To tell you the truth, I have been out of this work for so long I couldn't make an intelligent statement on that. The only thing I could say is that most of them are not going into research and design engineering, but into operational engineering. I think they do well [with] that. I think that is one reason why you are stressing management so much.

Rezak: **We certainly appreciate your sharing your time with us, Mr. Johnson, so that we could record some of your thoughts on the history of engineering technology education nationwide and also on the history of Southern College of Technology, of which we are also proud. And, quite frankly, the 10,000 or more graduates from the college and the 400-odd employees, of which I am one, wouldn't have the situation that we do if it hadn't been for your foresight, your leadership, and your tenacity in getting the whole project off the ground. So you are an important part of our tradition, and we appreciate your spending so much of your time with us today.**

Johnson: Well, thank you very much. I think the finest moments of my life were finding and developing Southern College of Technology. I am very proud of it.

Michael C. Mazzola

Former President Franklin Institute of Boston

Interviewed by
Richard P. D'Onofrio
President
Franklin Institute
of Boston

Born in East Boston, Massachusetts, Michael Mazzola began his career in technology as a welder. After serving in the Pacific theater during World War II as an aircraft navigation calibration technician, he attended Franklin Technical Institute in 1946 and graduated with a certificate in structural design and architecture (1948). He entered the Graduate School of Harvard University and received his M.S. in civil engineering in 1951.

He began teaching in Franklin's Evening School in 1948 and became head of the Civil Engineering Department in 1954. Five years later, he became dean of the faculty, and, in 1975, he was named director. In 1981, Mr. Mazzola was named the institute's first president and served in that capacity until his retirement in 1990.

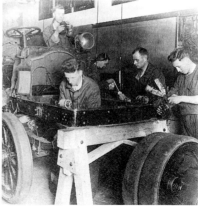

Oral Histories **213**

He has been an active member of both the American Society of Civil Engineers and the American Society for Engineering Education for many years and is a recipient of the James H. McGraw Award.

Richard D'Onofrio: **Mike, how, when, and where were you involved with engineering technology in ASEE?**

Michael Mazzola: My first involvement was somewhere around 1958 in the Committee on Conferences of the New England Section of the ASEE, and I served with Joe Marcus. At that point, we both had served on the Committee of 21, which was the original committee before the Technical Institute Division was founded, and we were the last members of that Committee of 21. And then in the early years, the 60's, the Technical Institute Division was formed, and we became a division of the committee, which was looking after the operations of ASEE.

Following that, the ASEE group formed the Technical College Council, and we moved up and served as vice-chairman and chairman of the Technical Institute Division. It was in the late 60's, early 70's, I guess in 1968, 1969, [that] I was chairman of the Technical Institute Division on the engineering technology study. And following that study I then, later on in the early 70's, was active in the implementing of that study.

D'Onofrio: **Was that the 1974 study?**

Mazzola: Yes. It came out in 1974. I was on that committee during all that time, from 1968, 1969 till 1974.

D'Onofrio: **What happened subsequent to 1974? What kind of activities and involvements did you have?**

Mazzola: The main involvement up to 1974 was the study and the revision itself. We were working on the revision and trying to get some activity going. The thing that I remember was that before the technology study, the four-year engineering technology degree came about. And there was quite a bit of controversy with all the engineering deans about the four-year degree. So this study was the result of that controversy.

D'Onofrio: Was it supposed to resolve that controversy?

Mazzola: Well, it was supposed to give it a little help, try to define it. It created quite a bit of friction between the engineering deans and the engineering technology people. Later on, when I became vice-chairman of the Technical College Council, and later the chairman of the council (I think that was 1977, 1978, and 1979), my activities were [to] try to repair the relationship between the engineering deans and the engineering technology deans.

I think that worked out very well. I had a very close working relationship with Dan Drucker and also with Bill Upthegrove. At that time that seemed to be the heaviest activity going on. There was still quite a bit of friction and controversy. But I think after that point we started looking at some government involvement and looking for some grants from the government for improvement studies. And I guess that, after I left, there were other studies that went on.

D'Onofrio: Michael, can you talk a little bit about some of the people who were involved early on, people like Walter Hughes and others who might have been involved in ASEE in its beginning, when it was called the Society for the Promotion of Engineering Education?

Mazzola: Yes. The Franklin Institute has been involved in activities in engineering technology or technical institute education, as it was called in those days, back before World War II. About the late 30's, early 40's, we had a Walter Hughes in admissions here; I guess he was some sort of a math teacher. He got involved with ASEE and worked with some of the early pioneers in technical institute education to get ASEE to recognize technical institute education as some part of engineering education, and Walter and that group, they were the ones that started a committee. They got ASEE to form a committee for technical institute education which later became the Committee of 21.

And following World War II, Walter Hughes went to McGraw-Hill and suggested that McGraw-Hill give some sort of recognition to the people in technical institute education, [since] McGraw-Hill was publishing many of the books that were used by these people. And the big moment came for McGraw-Hill. Walter McGraw, whom the McGraw-Hill Award is named for, passed away, and [another] McGraw decided to make the award in memory of this McGraw. I believe it was James [H.] McGraw. And they decided to make that award. Now, Walter Hughes died in 1948. He probably would have been one of the recipients, I am sure. But the first award was offered somewhere around 1950.

And then later on, Louis Dunham got involved. The next few people — you had the early pioneers — were involved in getting technical institute education recognized, and then we had over the next years Louis Dunham, Karl Werwath, Hugh McCallick, Dick Ungrodt, Cecil Tyrrell. This group, and Walter Hartung particularly, worked very hard with the engineering group to get recognition for the council and that is how the Technical College Council was formed.

D'Onofrio: **Can you tell me, Mike, what in your opinion were the most significant issues and activities in engineering technology that ASEE dealt with?**

Mazzola: Well, I think they finally recognized the need for engineering technology education, or technical institute education, as it was known at that time.

One of the engineering reports [was] the Grinter Report, on engineering and engineering technology. At that time, the committee which wrote this report made a division between engineering science and engineering, bifurcation as it was called. And what happened was because of Sputnik in 1957; they went engineering science and everybody forgot about engineering. And the technical institute group, engineering technology, jumped into the gap. And this is why the four-year program was started, because the engineering colleges were not doing engineering; they were putting too much emphasis on science.

D'Onofrio: **Is this where the concept of applied engineering came in?**

Mazzola: Well, that came later. This is the younger group now bringing out the idea that this is applied engineering, which is really what it was. It was the engineering that was done in the 50's and 60's.

D'Onofrio: **What progress do you feel engineering technology has made?**

Mazzola: I think that the fact that it did fill the gap and provided people to do the engineering applications which were not being done by the people coming out of the engineering science programs. There was too much research going on, and they were very important and necessary to industry. But someone had to do the scut work, the engineering practice, calculations, and the applications. And I think many of our graduates, all over the country, filled the gap and moved up into the position of management, and more knowledgeable people worked with applications rather than in research.

D'Onofrio: Mike, what in engineering technology's development did you not anticipate to occur over the years?

Mazzola: This is the concern I always had. I'm an old fuddy-duddy. That the engineering studies, I think, that the council replaced the engineering science; we tried to do what the engineering schools were doing instead of filling the gap. They had their thing to do. I think we should have done everything up to the point we didn't do anything. There will always be an overlap. We shouldn't be saying we are doing engineering when we were doing engineering technology.

D'Onofrio: In your opinion, what is engineering technology today compared to the 50's and 60's? I think you have already alluded to that.

Mazzola: Basically it is the same type of education. Technical institute education of that time was with the people who did scut work in the offices; they did the calculations, the applications, and worked under engineers. But now I think there is a gap between the engineers and all the graduates of engineering science and research people and some of the engineers; those are the ones that our people were working with.

In many instances now, the engineering technology graduates are the engineers. But they are the managers, engineering management, in many cases, a very necessary part of the spectrum of engineering.

D'Onofrio: If you had the chance, Mike, to share what you feel in terms of where engineering technology should be going and your own personal views with people who are involved in it today, looking back over the 45 years that you have been involved with it, what kind of advice would you give people with regard to engineering technology education?

Mazzola: One of the things I've always said [is] that we should be teaching engineering and engineering science if we want to. But we shouldn't brag about it. Keep your mouth shut and just get the jobs. You just antagonize the engineering deans when you start talking about it. But you just do it, and they'll give you recognition.

We had a wonderful relationship [in] New England with all the engineering deans. I was always treated as an equal. But you just did your thing, and you used to get registered. And two-year people used to get registered in the states, nearly all of them. The hullabaloo came up because they wanted to put it on paper and make it official. We always got registered. Once they made it official, we couldn't get registered until we got a bachelor's degree. So I think, do as much as you want, keep it low key, and just let your graduates go out and give you recognition.

D'Onofrio: Thank you very much, Mike.

Hugh E. McCallick

Former Dean
College of Technology
University of Houston

Interviewed by
Carole E. Goodson, Chair
Department of Civil,
Mechanical and
Related Technologies
University of Houston

Hugh E. McCallick is dean and professor emeritus of the College of Technology at the University of Houston. As a technology educator for over 40 years, he was instrumental in the establishment of the four-year degree in engineering technology.

During his tenure, he received numerous awards, including the Interprofessional Cooperation Award (Society of Manufacturing Engineers), the James H. McGraw Award and the Chester F. Carlson Award (American Society for Engineering Education), and the Commemorota Medalla (Universidade Federal de Santa Maria, Brazil). In addition, he is a Fellow of ABET, a life member of Phi Kappa Phi, a

Oral Histories **221**

member of Omicron Delta Kappa, and was elected vice-chairman of the World Engineering Congress in India.

He is cited in *Who's Who in Engineering, Who's Who in College and University Administration, Who's Who in the Southwest, Who's Who in America,* and *Who's Who in the World.*

Carole Goodson: Dean McCallick, thank you for meeting with me. Perhaps you could clarify with me when you started at the University of Houston: was it in the 40's?

Hugh McCallick: Probably. It must have been 1945...46.

Goodson: Was that your first entry [into] technology education?

McCallick: Yes, it was. As I recall, at the time I came to the University, technology per se did not exist. As a result of the war effort, specifically in what is now the College of Technology, there had been a number of industrial classes organized, such as drafting, machine shop, and then courses resulting from the Navy program: naval aviation, radio operation, and diesel engines.

So all this equipment was left, and there were classes still continuing. We found that as veterans came back, they wanted to improve some of their abilities in machine shop, welding. Aircraft was an important thing. Aviation was coming on; there had already been a pilot training program there. And when I came, the veterans started coming in to take some of these individual courses in mechanical arts, welding, drafting. The radio code course developed into a radio repair program. Diesel had moved from diesel mechanics to auto mechanics that included automobile mechanics repair.

So we had a cluster of little courses, and this led to no completion credential whatsoever. They took the course for no credit. It was then a part of the Department of Engineering, which was an amorphous sort of thing; nobody knew what it was at the time. Ray Sims, the first dean, had come from Reed Roller Bit to head up this particular unit. He had been teaching some industrial classes at Reed Roller Bit, namely teaching supervisory people the elements of supervision and teaching how to teach.

Goodson: Training?

McCallick: Yes, training. It was called TWI, Training within Industry. And I joined him. He asked me to teach some of the work in aircraft.

Goodson: So you had known him? He inspired you to get involved?

McCallick: Well, what got me was Professor James Hutchinson in mathematics. He had run the aviation training program. He was the director of the aviation program, and I taught navigation for him so he asked me to come. And when they brought Sims, I stayed.

Then it occurred to me, and this, I think, is a significant thing. It is significant because the programs started at the University of Houston were patterned after no other program anywhere.

Goodson: At the University of Houston?

McCallick: At the University of Houston. There wasn't anything like it anywhere.

It seemed to us that students, the veterans coming through, had a problem becoming supervisors. So they needed more vocational skills, and they needed a little more arithmetic. I wrote the first arithmetic book. It consisted of 21 pages of arithmetic: addition, subtraction, fractions, and decimals, and we handed those to the students.

Goodson: Things never change.

McCallick: And I started seeing an encouraging sign among faculty; the faculty came from all over the campus. Faculty from engineering, English, [and] what is now the College of Education came to teach classes. And people from industry came to teach classes. There was a "polygogden" of backgrounds which fostered a lot of talk and a lot of thinking. We came up with the idea that we ought to award these students some sort of completion credential. So the first was a one-semester program which involved machine shop, and a person would take arithmetic and some drafting, and we started to cluster these several courses into a program that was keyed to the certificate of completion.

Then radio repair bifurcated. We had a teacher named M. Masser who insisted that the field of radio should be angled very strongly towards communications rather than repair, and so he started to require algebra, which was unthinkable. Algebra: can you imagine? It became a nine-month program for which we awarded the Certificate of Achievement, to try to differentiate the two. These programs evolved in clusters.

Finally, programs seemed to separate into two general areas: one vocational and the other more technical, i.e., a diesel area, a machine shop area, an automotive area, a radio repair and television (which started about that time) area, [which] later developed into what we called electronics. We picked that name out of the sky. Transistors had not come on the scene; it was all vacuum tubes. Drafting became more sophisticated, requiring more math and reports.

As the responsibilities of the people we produced increased, they would come back to school. In time, there evolved two distinct groups, one more manual arts-oriented [and] one that had a larger segment of so-called theory. It involved English and report writing, algebra rather than arithmetic, and then the unthinkable: we added trig.

A teacher by the name of K. W. Feist was responsible for supplying a lot of surplus equipment. He taught radio code for the Navy aviators and others, and Feist studied in math and evolved a very high proficiency in using and teaching the slide rule.

Goodson: **We still have that model.**

McCallick: It's an artifact. It was a problem to get that slide rule because it didn't belong in a manual arts program.

Goodson: **It was more for engineering?**

McCallick: It was an engineering tool. So Dean Ray Sims and I decided (and this was a joint decision) to divide into the Schools of Technology, with a vocational and a technical division. And then we had a problem with the faculty. They were not acceptable to the main stream designations and ranks. We had our own graduation, of course. They didn't participate in the chapter U of H graduation at that time.

Goodson: **The traditional university graduation.**

McCallick: Right. So we evolved a dual ranking and called the people in the vocational or manual arts programs "technical instructors," and we differentiated by calling the others "senior technical instructors." We made a point that the senior technical instructors were comparable in rank and pay to assistant professors. Therefore, the requirements and qualifications started to increase.

Goodson: **At that point, there was no degree requirement. Was it experience or...?**

McCallick: No. It was at that time, the only doctorate was a doctorate of divinity. Probably 90% of the faculty did not have a degree. You were an exception if you had a baccalaureate degree.

Goodson: **Not even a bachelor's degree?**

McCallick: A bachelor's, no. This was amazing. We encouraged faculty to go to school and get a degree. I think of Dr. John Martin, Dr. B. C. Kirklin, William Roulette, McDonald, and some you did not know of.

Probably the most significant thing that occurred at that time was in the late 40's. We decided (at that time the university had a junior college division, a municipal junior college) to teach a lot of the tech courses through the junior college structure, and we got state support for that division, even though the UH was a private institution. So we said, "Let's see if we can't fit into the package and mold a little better." And we came up with the Associate's in Applied Science. So we really had an upper- and lower-division university.

Goodson: **And it went through the junior college division.**

McCallick: It went through the junior college division. So this was one of the significant issues, to get acceptance of the courses leading to the certificate of completion to apply towards an Associate of Applied Science which [was] a new degree. This was totally new. It didn't exist.

Goodson: **Do you mean that no place else in the country had this degree?**

McCallick: No place else that I know about had an Associate of Applied Science. They had an Associate of Arts, an associate in other things, but not applied science. So we started it and sent it through, and it evolved.

Then we found that our associate programs were three years in length, and they led only to an Associate in Applied Science. About that time the college was strong in distinctly vocational arts programs, meaning automotive, welding, upholstery; we taught a lot of the pipeline welder's certification programs, radio repair, machine shop, and certified apprenticeship-type programs. And then the others evolved into a civil program which was not yet totally construction-oriented. We involved faculty from all over the place, and we started the "related courses" area. We didn't know what to call it; it was unique.

Goodson: **It still is unique.**

McCallick: That provided us with a control over what was taught in those courses; that's when we developed technical report writing.

We had these programs, in terms of effort, four years in length. So I thought, "Well, let's see if we can put these things together into a program that would lead to a bachelor's degree." Everybody's hair went straight up on this. I remember talking to the dean, and he said, "Let's talk to the president about it."

We came up with the term "Bachelor's of Applied Science." I borrowed that term from a program that I saw in England. I called it a BAS so it wouldn't compete with any other program. I asked a number of liberal arts faculty to help develop the program. We talked with the president, and he said, "Develop one." I remember walking down the hall trying to talk to him about it, and he said, "Mac, go ahead try it; let's see what it looks like." That's how easy it was to do things in those days. Can you imagine: no committees, no councils?

So I asked a number of the faculty, a distinguished French professor, an English professor; we had about seven or eight impeccable academic leaders at the university come together. We said this program has to satisfy two objectives: number one, to the best of our ability, the program must assure the technical capability of the graduates; number two, it has to satisfy all the requirements for the bachelor's degree. We came out with a program that required 152 hours to complete. We don't have that in programs today.

The first BAS was awarded in 1952 to a student in civil technology; the term ET was not applicable.

Then I left for a number of years, and we had problems coming into the state system; a number of changes were forced. They were forced to give up the Bachelor of Applied Science degree.

I think that degree was one of the most significant events in technology.

Goodson: Because it was new?

McCallick: Mainly, Carole, because it gave opportunity. It opened the door for a student to be in a position that did not delimit his educational process. He could go from an associate degree to a bachelor's and, in time, a master's and even a doctorate. It provided a path, an opening.

Goodson: And I guess that at that time, associate degrees were not such that people moved freely from the associate's to the bachelor's degree in any discipline?

McCallick: Oh no. In fact, very few people got an associate's degree.

Goodson: Were there other two-year technology degrees back in that time period?

McCallick: Well, they had started some two-year [programs], but the leadership in what we now call ET was provided by a group of 10 or 11 private institutions; they were not public institutions. These institutions were Wentworth, Ohio College of Applied Science, the Academy of Aeronautics under Casey Jones, Capitol Trades, Chicago Technical College, Aeronautical University, Milwaukee School of Engineering.

Goodson: And these institutions were two-year degree programs, or were they more like technical training programs?

McCallick: Well, no. They were 18 months long, in effect three semesters and two years; two years that in our terms are not two years but continuous-type programs. Private institutions provided leadership to ET not evident in public institutions.

Goodson: Were they like clock-hour programs, with no degree?

McCallick: They didn't give a degree; they awarded a certificate of completion or something. They were not approved to give a degree.

Goodson: **Would you say that they are analogous to the clock-hour programs that exist in community colleges today?**

McCallick: Exactly. Clock hours. Many of them were geared to returning veterans. The government paid tuition on a clock-hour basis rather than a semester-credit hour basis.

Goodson: **And some of them were associated with universities and others were not?**

McCallick: No. At that time there were not any, that I know of, that were associated with universities, except ours. There may have been others that I don't know of. We were in the middle of the technology movement from private to public, including the community colleges.

Goodson: **So it was a unique situation since UH was both public and private in a sense?**

McCallick: Very unique. And this segment called technology had been growing tremendously and was extremely strong. Remember during the war, private industrialists, including H. R. Cullen, Francis Berleth from Hughes Tool, Gulf Publishing — there were some 21 companies in 1938 that built our manufacturing building, the third building on campus. They put it together in 1939 as a manufacturing building; it cost $250,000. Cullen underwrote it. He guaranteed the money, and the money was actually raised from these 21 different industries.

About that time the war hit, so it was never used initially for the original purpose: to develop programs that would benefit the business and industrial need of the Houston community. That was the purpose of the technology building; that's why they provided the money. It was used for war purposes and later developed into the College of Technology.

Goodson: **Which really did benefit those industries?**

McCallick: I think it is fascinating that during the war those programs supported the university during very hard times.

Goodson: So it was that training that kept the rest of the operation going?

McCallick: It kept everything going. There was a different ambience.

Goodson: A different attitude toward technology on the part of the faculty?

McCallick: The professors that had been there lived through this period and knew the place within the structure, and it was very viable. And when they got state support, they eliminated the area. That's when I came back.

Goodson: Was that in the late 50's, early 60's?

McCallick: They eliminated it in '60, and the program went back to the associate degree. When I got back there, I think we had 96 students and a handful of faculty: G. McKay, J. Martin, B. C. Kirklin.

I came back mainly to get back to business in Edna, and I came back to run the program. Then the dean died, and they asked me to be dean of the college. By 1962, there were a number of technology programs around the country.

Goodson: Do you mean programs like the one at Purdue?

McCallick: Well, we probably had the first.

Goodson: With the Bachelor of Applied Science?

McCallick: But that disappeared. And there were a number of programs, and they were trying to do something about accreditation. So I undertook a study which became known as the "McCallick Report." This report started looking at programs, to come up with guidelines. The committee was broad-based, including the dean of engineering, Pete Lohmann from Oklahoma State, and others. It was a very strong committee.

Goodson: What organization was this under, ASEE or ECPD?

McCallick: ECPD. And they asked me (at the time, I think, Larry Johnson from Georgia Tech was the chair of the sub-committee dealing with the review of two-year programs in ET) to chair the group, and I did. It included the current head of engineering and science programs from the U.S. Office of Education as a member of the committee.

If you can imagine the debates around the country about developing guidelines for four-year programs in ET, I debated and had tremendous discussions with engineering deans and others all around this country.

Goodson: I guess they weren't really thrilled with the idea.

McCallick: They weren't exactly ecstatic. When I finally came and wrote the guidelines, and I want to stress that — guidelines for the evaluation of technology programs in engineering — we developed a rather extensive report, widely accepted. It was called the "Bible" for the evaluation of programs in ET; that was what it was termed by people who quoted it frequently. The "McCallick Report": it became part of the literature.

It was then used to approve programs. We were the third institution to be reviewed. Purdue was the first, and I chaired the committee reviewing the Brigham Young program. The people on the committee who reviewed BYU consisted of the chair of the Engineering Accreditation Committee and four or five deans of engineering around the country, plus one or two others from technology. I felt that whatever happened to this program, it had to be without question. If it [were] approved, it had to be strong. That's how we approached it. And then, of course, we came up for review at UH, and it was approved.

Interestingly enough, technology to me meant a field of activity, related to fields in other professions; more than engineering. That's why I always insisted that we wanted a bachelor of technology degree. Under this umbrella, we could evolve all sorts of applications-oriented programs, and the guidelines we developed were guidelines.

What has happened is that the guidelines have become standards. [But] standards are what has happened in the past; guidelines look to the future. And if anything has been wrong with engineering technology and the accreditation process, [it's] that evaluators became more interested in applying standards which tended to stultify the experimentation of faculty with new programs, new ideas, new directions in this field.

I don't think ET has moved as rapidly as I hoped it would, and if I were to attribute it to something (there are probably a lot of reasons), [it is] that standards, rather than guidelines, have been used, and they have curtailed faculty. ET has not assumed its responsibilities as fast as I hoped it would by now.

Goodson: **What do you see as other factors?**

McCallick: Well, one other factor is that, for some reason, people in ET have been unwilling or unable to develop their own criteria for the evaluation of faculty. As a result, there has been foisted on this group evaluation systems that are not relevant to this field of work. As a result of trying to be all things to all people, rather than one thing to ET, faculty have tended to disperse their energies in areas that are not really significant.

Goodson: **And perhaps some of the new ones have no understanding of that because they are into the system?**

McCallick: They see the system: that's the system; that's what you do, and that's what you're going to be judged by.

Goodson: **And they tend to advocate it....**

McCallick: Well, of course they would. There is no question about it: "Who are you to be different?"

Whereas with ET, what flourished once was a willingness to imagine, to be different, not simply to be different but to fit needs. I think that ET has gotten out of whack with the requirements of jobs and with the future.

Goodson: **That's interesting. Do you mean generally, totally, in all areas?**

McCallick: I think in all areas. I think it is out of kilter. It's behind, not ahead. The requirements are generally ahead.

It takes imaginative faculty, not industrialists, to be able to anticipate and to grasp developments and interpret them into educational programs. Now if you're forcing this faculty member to comply with ABET requirements and the requirements of the program, you have a real problem. What can you do? So I think this has tended to stifle the kind of growth that I had hoped would occur.

Goodson: **Do you think ET educators should work to change the system that they are in or do they need to move out of the system into another?**

McCallick: Of course, the easy answer would be to separate the institutions, one of a separate entity from engineering or anything else.

In fact, engineering is moving more into the areas of technology. The move in the bachelor's programs in engineering [has] been very definitely in that direction. That's fine. Somebody has to fill the need; whether it be this or that doesn't make too much difference. But the institutions that have done best are stand-alone institutions.

No one is going to move technology forward but faculty, and I'm not sure that they have the reward system. It is becoming more stratified. They may not be of the right mixture to move it. It takes great individual leadership; it takes a willingness to take a chance.

Goodson: **There are a couple of issues facing technology today. Some technology educators don't think that we should be technology per se; rather we should be accredited under the name of applied engineering. Another issue centers around registration of technologists as engineers, an issue that's been around off and on for a long time, but perhaps emphasized more now.**

McCallick: Well, I have not been involved with it recently, and don't really know. Philosophically, my position has been, is, and, I suspect, always will be [that] you get nowhere riding somebody's coattails. You simply don't get there first, riding someone else's coattails, if you want to get there first.

Maybe that's the way it ought to be; I don't really know. It would be interesting for some people to look at, but I think the accreditation process needs to be either eliminated or totally changed.

Number two: I think that you need to take a good look at what's really happening, not only in this country but all over the world in the field of technology.

Goodson: Such as in Australia, where they have just established a program in technology?

McCallick: I was there. We have in this country been too willing to accept the criteria that has been set up for others. You have to develop, imagine, dream your own. What is it that you want to do? How do you get there? Then develop ways, guidelines, directions. You can't apply standards. Standards are past; standards are comfortable.

Goodson: So you think that ABET criteria should be revised as guidelines?

McCallick: Yes, and I would say that about engineering also. Look at the battle relating to accreditation of graduate programs in engineering. It's been fought for years, and there are still few.

Now regarding registration, it's a wonderful thing, but it does not yet mean certification in the same legal way that it does in the profession of medicine or the legal profession. There is not the same impact, not that there should be. I'm just saying that there has not been. You're fighting the wrong battle, I'm talking about in technology.

The other aspect of technology is that if you accept technology as a broad form of activity, which encompasses many fields, it's too bad that you've missed some of the opportunity to develop this area of higher ongoing education.

Goodson: You mean graduate programs?

McCallick: Why not? Is education terminal? How do you provide continuing recognition for people who continue to grow educationally?

Goodson: I think it's coming but it has not been rampant. And there are those, mostly in engineering, who don't believe it is appropriate.

McCallick: Remember the business program at Harvard? There was a tremendous battle about it; it's not appropriate at an institution of that sort! And now when you think of Harvard, what do you think about?

Goodson: **Are there some fields of technology that you believe should be developed that are not being exploited at this time, some that you see emerging?**

McCallick: Let me approach it from a different point of view, accepting the fact that technology deals with technique, applications, etc.

I was in Spain last week, and the program that covers Europe is CNN. The most important thing we have is information. We have it to disperse, dispense, and sell. Information does not necessarily mean a written, theoretical thing. It can be tangible, but it encompasses a huge area that is applications-oriented, and it provides an area of development for technology. When you think of information, you think of computers, but there are other concepts, many.

Information is a broad concept. There are two things I see around the world: one is information (we seem to be the best dispensers of information), and the other is design of consumer products.

George W. McNelly

Former Dean
School of Technology
Purdue University

Interviewed by
Michael T. O'Hair
Administrator
Purdue University
Programs
at
Kokomo

George McNelly received a B.A. in chemistry from Coe College (1950), an M.S. in applied psychology from Iowa State (1952), and a Ph.D. in industrial psychology from Purdue University (1954).

After 26 months as an electronics technician in the Navy during World War II, he worked for Inland Steel Company for eight years in supervisory positions. He holds 12 patent disclosures and two patents related to protective coatings and steel-making processes.

Dr. McNelly became dean of Purdue University's School of Technology at the West Lafayette campus in 1966, with responsibility for 3,000 students, 135 faculty, 8 departments, and 20 degree programs, ranging from associate

to doctorate at the main campus and eight regional locations. He also spent 10 years as an educational consultant at Kabul University, Afghanistan, and was one of two U.S. members of the Steering Committee for the World Conference on Engineering and Engineering Technology, held in Köln, Germany, in April 1984.

Michael O'Hair: The purpose for this taping is to record oral history pertinent to engineering technology education related to the American Society for Engineering Education.

Our first question today, Dr. McNelly: could you tell me about your academic and industrial experience that prepared you for leadership in the development of engineering technology at Purdue University?

George McNelly: Mike, my background is what I'd call a little bit about everything. I have a general background. I started in technology in 1944, when I joined the Navy and became an electronics technician. I went through what was called then the EDDY program. It was an intensive year's training in modern electronics, and it really was equivalent to probably two years in electrical engineering because of the intensity of the program.

We spent a month in Chicago in what was called pre-radio and then three months [in] primary radio activity at Oklahoma A & M, as it was known in those days, in the electrical engineering department. We then spent the rest of the time, something like eight months, at Navy Pier in Chicago, where we had advanced training in just about every kind of electronic equipment that the Navy had at that time (sonar, radar, all kinds of advanced communications equipment). I spent a little over a year out in the field assigned to an LST in the South Pacific. I was on an LST when the war ended.

O'Hair: For those who are uninformed, what is an LST?

McNelly: An LST is a Landing Ship Tank. They were vessels that were not too big, but they carried tanks for the Marines and would land the Marines on islands because they would actually go up on a beach and then they would pull back and become hospital ships.

I was the electronic technician on one of these, which [was] an interesting experience. I was in the Navy about 27 months, and after that I came back to the United States and went into a place called Coe College in Cedar Rapids, Iowa, a small liberal arts college. I wanted to go to Iowa State and major in chemical engineering, but Iowa State had a lot of veterans back at the time and they said, "You could come, but you're going to have to spend the first year in Des Moines at an old Army camp."

I thought I had enough of the military and that kind of life, so I went to Coe College in the town where I lived. A very fine school, by the way. I majored in chemistry, mathematics, and physics and enjoyed that kind of program very, very much. I also took many liberal arts courses, and my senior year I took a course in psychology and really enjoyed it.

When it came time to graduate, I wanted to go on to graduate school in chemistry, and I had several offers of assistantships around the country. But they didn't pay an awful lot and I was going to get married. My psychology professor said, "Psychology is a natural for you. I can get you an assistantship that is better than you can get in chemistry." So he made a call to Iowa State and talked to the head of the department, and I immediately got an offer to have a graduate assistantship there in the experimental psychology area, and I said, "Why not? It's more interesting than chemistry." So I switched fields my senior year, having had only a couple courses in psychology, and so I graduated in chemistry from Coe and went on to graduate in psychology at Iowa State.

I started to teach psychology and taught the first course I had a year after I took it. That's the way to really learn a field. I also did work in the experimental lab making equipment. This was a two-year master's degree program in applied psychology, and during this time there was a visiting instructor from Purdue who said, "You want to come to Purdue and be an industrial psychologist."

So he took me to Purdue and I met the key people. The main ones, of course, were Drs. Lawshe, McCormick, Kepart, and Tiffin. These were the men [who] were running the finest industrial psychology department in the country. They gave me an offer of an assistantship which provided full support for doing my doctorate. I came to Purdue with my doctor's thesis idea, and they assigned me to Ernie McCormick, who was probably the nation's finest person in human factors and design of equipment. I learned a great deal at Purdue, and my second year I became a full-time instructor as I finished up my thesis.

When I graduated in 1954, I was given an opportunity to stay on as an assistant professor. I couldn't afford to do it; my wife was pregnant, my car was worn out, and the place where we were living was ready to be torn down. I added it all up and found that I couldn't afford the great opportunity to stay at Purdue. I was given another opportunity to go to the Inland Steel Company to take what was called a "loop program." My introduction to Inland was through Maurey Grainey. I took a course from him. He had been at Inland the previous year starting a training department. Maurey Grainey later went on to the University of Dayton and became the dean of engineering.

O'Hair: **For the record, where is Inland Steel located?**

McNelly: Inland Steel's main office is in Chicago, but their main steel-making plant is in East Chicago, Indiana.

O'Hair: **So, you remained in the State of Indiana?**

McNelly: Yes, I stayed in the State of Indiana. I was introduced to the people up there because Maurey Grainey took us on a field trip, and I met a number of the executives, and somehow or another one seemed to be impressed enough to give me the offer to be a member of this very elite small group of people who would take this loop course. That was about a year-long program. It was an introduction to every aspect of steel-making. At the same time, at night, we would take metallurgy courses and things like this. Although my chemical background was ideal, I still took the course.

Unfortunately, the first night I was in this particular metallurgy course, Millard Gyte, who was the director of the Purdue Calumet campus, came into the classroom and said, "You've got to come and teach another course for us." So I was pulled from that class. I said, "I've got to stay here," and he said, "Oh I've already talked to the instructor; you can just take the exam and read the textbook." So my first brush with metallurgy was reading the textbook and taking the exam. One of the other instructors had gotten ill, so I was teaching her courses. I taught at night at the Purdue Hammond campus starting in 1954, and I continued to do this almost continually through my eight-year career at Inland Steel Company.

The first year of introduction to steel processing and every aspect of steel-making was very, very interesting, and I found that while I had started at Inland thinking I wanted to get into staff work, industrial relations or something like that, I changed my mind. At the end of the program, I selected the galvanizing department, a very advanced technical department. I actually went to this department and spent seven years there in production. One of my jobs was to help design [and start] new facilities. I had a lot of work with engineering and science teams, as well as all kinds of electronic and mechanical equipment. This was really my introduction to the use of technicians in industry.

In 1962, Chuck Lawshe, who had become dean of regional campuses at Purdue, was looking for someone to head up their two-year associate degree programs. I had occasionally visited the Lafayette campus and run into him, and he gave me a job offer to leave industry and come back to Purdue. In the meantime, salaries at Purdue had come up to a point where you could live on what they were willing to pay, so my wife felt that it was time to leave industry. We came back to Purdue in the fall of 1962.

O'Hair: **Could you tell me about the early years of engineering technology development at Purdue, and your role in it?**

McNelly: What had happened, Mike, is that back in World War II, Purdue University and Indiana University started what became later the regional campus structure. These facilities were built so that people living in the major communities of the state could take more training and become better able to do the job in industry. Purdue's campuses were technical and were run by what was called the "technical institute."

The most important person in this regard was a man named Beese. I didn't know him well, but he was an industrial engineer, as I understand it, and was head of this technical institute activity. The man that had the concept and I guess got the War Training Act passed was, of course, our very illustrious A. A. Potter, who was dean of engineering at Purdue for so many years. He was a marvelous man who lived to be something like 95 or 96 years old. He was a very dear friend of mine in his later years.

Dean Potter is the one that got this War Training Act through in 1943. The idea was to train manpower so they could do industrial jobs needed to win World War II. After World War II, these centers were going to close and the people in these cities said, "No, leave them there. We need this kind of thing." And it was then that they set up the technical institute activity which Beese headed.

Centers were at Hammond, the Calumet campus, Fort Wayne, Indianapolis, and Michigan City. These were the take-off places for what [were] later to become full blown campuses. Indiana University had a similar structure, but not offering the technical work that Purdue did.

When I came on the scene in 1962, Beese had been dead for several years. A man by the name of Had Roundtree, who was an old-time technical institute leader, had come from the East Coast to lead up the program at Purdue, but nothing was happening. Chuck Lawshe had upgraded the old technical institute courses to give regular college credit and this resulted in an Associate Degree in Applied Science for two years' work.

O'Hair: **Was this then housed under a specific school?**

McNelly: This was located in the technical institute, which was housed under the regional campus administration headed by Lawshe. The faculty body that made changes in courses and curriculum were actually people representing the Regional Campus Council. I think it was called that, and these were representatives of each of the schools on the Lafayette campus.

This was, in a sense, the official faculty body. They had curriculum committees but they didn't really have an autonomous faculty. This amorphous council did all the faculty activity. There wasn't any real problem with this, and it certainly was a means of involving the rest of the campus, but there really wasn't the cohesion that one would want in a faculty. Lawshe was somewhat perturbed with Roundtree because Roundtree was very busy and didn't seem to have time to communicate.

Dr. Lawshe had grandiose plans as to what could happen in a two-year area. He felt that industry needed these kinds of people, particularly in the engineering technology area, and wanted me to come in, clean house, and get this thing off the ground. Well, I immediately realized that Roundtree was a valuable source of information and contact with particularly ASEE and the old Technical Institute Council. I felt that he was not being utilized properly, so I first of all made him a close friend, and I think that he did a nice job until he died. He was quite effective in his later years.

As soon as I came on board we had what was then known as [an] ECPD, now known as ABET, accreditation visit. This was in

September of 1962. We had all of our programs looked at, all of our engineering-related programs at each of the regional campuses, as well as Lafayette. Lafayette did not have these programs, only an aviation program at the time. They looked at that also, and about eight months or nine months later they gave us the decision that these programs were not accredited. In the meantime, they had changed the criteria for evaluating these programs, upgraded them as obviously they ought to, and so we had a whole new challenge.

By that time, I had my feet on the ground and knew what we needed to do. Of course, one of the major things was to reorganize and set up departments instead of having a Division of Engineering Technology. When the vice-president called and wanted to know what was going on here, he said, "What are your plans?" I said, "What we need to do is set up mechanical engineering technology, electrical and also civil engineering technology departments because these were three major thrusts in engineering-related technology."

The vice-president wanted to put money in this and, lo and behold, he put $100,000 into these programs. These were the days when vice-presidents apparently had a desk drawer where they kept cash. We made it go farther by obtaining dollars from the Vocational Act to match this money. Here was money that Paul Chenea, the vice-president, had given us, plus the money from the Vocational Education Act. We were able to get almost $250,000 in one fell swoop. Believe me, in 1963 that money went a long way.

This helped us get computers (IBM 1620's as I recall). Of course, that's really telling you my age, but we were able to get computers at Indianapolis, at Fort Wayne, at Calumet, and we were able to fully equip the electronics labs at each of those locations. We were able to get first class overnight, and this really helped because the faculty morale was low, and they thought maybe we were going to phase the whole thing out for not being accredited. The faculty morale went sky-high when we were able to do these things.

In the meantime, Chuck Lawshe, who was a great dreamer and a great planner and a great person to look ahead, felt that this faculty ought to be autonomous and ought to have their own school. He selected a committee made up of people from his Regional Campus Council and these people were asked to handle the challenge of developing a new school at Purdue University. I was on this committee. Ken Michaels, who was an associate dean for Lawshe and also an old-time buddy of mine from the psychology department, chaired this, and we came to the conclusion that a new school was certainly feasible and the Purdue school could be named the School of Technology.

They did their work in 1963 and recommended this to the Board of Trustees, and the School of Technology became a reality on July 1, 1964. This school included three elements that were already in existence: the Division of Applied Technology, which coordinated all the regional campus two-year technology programs; the Department of Technical Applied Arts, which was really part of Industrial Engineering (they were located in [the] Michael Golden Laboratories on the Lafayette campus and provided the lab courses for the industrial arts people); and Industrial Arts, a section of the Education Department. The Technical and Applied Arts was headed by Denver Sams, who later became my associate dean. What we did was poll the faculties in each of these three elements and simply said, "If we were to develop a school of technology, would you be willing to transfer to it?" We got 100% support.

The right chemicals were put together at just the right time. As a result, we grew at a tremendous rate. I don't think even a man with the vision of Lawshe thought we would grow as fast as we did. We started with a total of about 1,200 students, full- and part-time, at Lafayette and these various regional campuses in all programs that were put together as the nucleus of the School of Technology. We ended up at the end of a decade having probably 12,000 students pursuing degrees at all campuses, full- and part-time. We must also remember that this initial group also included a nursing department. You might wonder why, but at the time the Board of Trustees said that we were to coordinate all two-year programs, and nursing started as an associate degree program. It later evolved into what we now call the two-plus-two program.

Of course later, in 1979, nursing left the School of Technology to become a school of their own under the schools of pharmacy, nursing, and health sciences. They had some 600 students, and they've evolved into what would best be described as a generic bachelor's program in nursing. It is no longer a two-plus-two program.

At any rate, the School of Technology was unique in that we initially saw a need to sense where the problems were, where the needs were in the state, and then do something about it. We started out with a sensing element called the Office of Manpower Studies which was directed by J. P. Lisack. I chaired the committee that set up this kind of mechanism, and we were the committee that found J. P., who did a fantastic job. What it meant was that this office would initiate manpower studies working with industry and interested people in various communities to uncover the needs. If we were to start a program, what would the program be? Under J. P.'s leadership, the Office of Manpower Studies documented the needs and turned out a manpower study. That was a tremendous mechanism for involvement.

Our whole school really was a mechanism for involvement. The reacting element then became our faculty and our laboratories. Our curriculum was developed in ways that could be flexible. Follow-up was accomplished through follow-up of the graduates and working with business and industry to find out how they were doing and what the feedback ought to be in terms of changes of courses and curricula. We did this through an industrial advisory committee.

I think this approach was unique. It provided us with a strong backing from business and industry, because in the manpower study involvement we would uncover needs that each of these industries had, not only at the moment but in the future. This meant that we had ready information about where to place our graduates, and, as a result, we never had any difficulty in placing our graduates. The moment we had this record of placing all of our graduates, it acted as a stimulus for more students to come to the school. We were getting more and more students and placing them well. This was the secret, if there be any secret of success.

O'Hair: **Would you focus for a moment specifically on the engineering technology programs, EET and MET, a little bit about some of the challenges, the personnel that were involved? Also, would you like to talk about why Purdue doesn't have a civil engineering technology program?**

McNelly: We came in with the suggestion that we needed to reorganize into departments. We were very fortunate in having already on board, as kind of a diplomat without portfolio, a man named Gil Rainey. Gil was one of the early pioneers and great leaders in the field of electronics at the technical institute level. He was coordinating these programs but really didn't have a title. It became a natural to make him our first head of what became electrical engineering technology.

He led that area for a decade and did it in a brilliant manner and made this thing grow like wildfire.

In finding somebody to head the mechanical area, we had a more difficult problem, but, fortunately, an old Purdue grad walked in one day and indicated he'd like to come with the academic. His name was Walt Thomas. Walt had been down in Florida working with Pratt & Whitney. He was the ideal person to head up our first mechanical engineering technology department.

Initially, the electronics was in Indianapolis; we didn't have engineering technology on the Lafayette campus, and so Gil Rainey centered his activity there and coordinated all the other campuses. Walt Thomas was up in Fort Wayne where he headed up the program there and, of course, coordinated this activity at all the other campuses. It was, I think, two years or so before we got engineering technology on the Lafayette campus.

In the meantime, this problem of civil engineering technology was [further complicated] by having architectural engineering technology at Fort Wayne, civil engineering technology at Calumet and also at Indianapolis. The Calumet campus program was headed by a man named Chuck Hutton, who was a brilliant architect. He did some magnificent things, and his students were all practically oriented. He was an old-time person in this area, even though young in years. We made him our first head of this area. I think we called it "architectural and civil engineering technology." Chuck Hutton gave us leadership in this area for several years and again centered his activity at the Hammond Calumet campus.

When we were ready to bring this into the Lafayette campus, Chuck was not about to leave Hammond. This was his home. His family ran a big construction company up there, and he loved teaching and really didn't want to continue on as head. So we appointed Dr. Dorsey Moss as head of this area. He was at the Fort Wayne campus and headed it up there for awhile. We brought him eventually to the Lafayette campus, where he gave outstanding leadership to this program for a long time. It eventually became construction technology.

Each of the regional campuses, particularly the Fort Wayne campus and Indianapolis campus, became large. The programs became large, and, in some cases, there were up to 3,000 students in some of these areas on these regional campuses. It was not logical to center the leadership in Lafayette. So like an amoeba dividing, as the campuses reached a critical level, we gave them academic autonomy. The Fort Wayne, Hammond, and Indianapolis campuses became academically autonomous. They had representatives on our council of representatives, which was a faculty body much like a senate. As far as leadership is concerned in the school, Chuck Lawshe was dean of the School of Technology at its outset from 1964 to 1966, and I was the associate dean. Actually, the truth is that he had many other things to do with regional campus activity. I reported to him, and I was in charge of the School of Technology. At the end of two years, I was made officially dean of the school and was dean of the school for 23 years.

It was an exciting 23 years, and the time passed so quickly because we were in a growth mode. At all times we were challenged to come up with up-to-date laboratory equipment, and our friends in industry clearly made the school what it is today with all the gifts and support they gave us, not only directly but indirectly in the legislature. Of course, along the line we built buildings and had better-equipped laboratories everywhere, and we also broadened our activities through the mechanism called statewide technology, with Kokomo the largest and clearly the crown jewel. Whatever you want to call it, it is the largest and best of all of these other campuses.

We opened up technology opportunity on the various Indiana University campuses, and so today we are scattered all over the state. I think we very effectively bring engineering-related programs to the areas where they are needed in the state. The School of Technology is more than engineering technology or engineering-related technology. The two major engineering technologies are mechanical technology and also electrical. But we have, of course, an aviation program, and this includes both piloting and administration and the mechanics. We also have a computer technology area and a supervision area, and the industrial arts area has become industrial technology.

Of course, we have building construction technology, and we now have technical graphics which used to be engineering graphics. Some seven years ago, engineering graphics was going to be in our new building and was transferred to us from engineering. Today we've got eight departments including computer technology, which really is systems analysis and things like this and not the hardware of computers. This makes up what is known as the School of Technology.

O'Hair: OK. I would like now to focus on ASEE, with this background of evolution of not only engineering technology, but technology at Purdue. The first question I would like to ask is, how and when were you involved in engineering technology within ASEE, and could you tell me a little bit about the people you recall and the connections there?

McNelly: When I became director of this Division of Applied Technology, which included all the engineering-related technology, I joined ASEE. This would have been about October of 1962. Through Had Roundtree, I was introduced to what I would call the old-time technical institute group, later becoming engineering technology leaders: people like Karl Werwath; Ken Holderman; Russell Beatty, who headed up Wentworth at the time; Larry Johnson from Georgia Tech, and, as you will recall, their technical institute was part of the engineering group; Ray Sims; Curly Foster; Walter Hartung; Winston Purvine; Hugh McCallick, who was with the University of Houston; Mel Lohmann and, of course, Dick Ungrodt, who was an old-time leader in this area; Ross Henninger; Merritt Williamson; Don Metz and Joe Gershon; and Robert Wear. These were some of the people that Roundtree introduced me to that were leaders in the field, and all of them became friends.

In terms of ASEE itself, I went to my first meeting in June of 1963 and gave a paper. This was in Philadelphia, and as I recall I had a paper that was describing the differences between engineering and engineering technology. So as far back as then, we were concerned about this problem of becoming unique but still related to engineering. We set up our departmental structure with Thomas and Rainey heading two major areas.

It became rather clear, particularly with the kind of budgets we were dealing with, that it was nonsense for me to go to ASEE meetings, particularly if they were not within driving distance, but it was essential for the heads of these two engineering technology departments to go and become active leaders in the field. While I had a lot of knowledge of the people I just mentioned (they were the key leaders, early leaders in this whole movement), my spot was to pretty much stay at home and send the important people.

My involvement in ASEE has always been peripheral. I would go to an occasional meeting, like once every five years, and meet with people, but my major thrust was to meet with some of these leaders when they came to visit our laboratories to see what Purdue was doing. We made certain the mechanical and electrical people were active, and we would always support them a couple of meetings a year anyway. As much as possible, we sent faculty members to ASEE meetings. I can't say that I was active in ASEE activities in a direct way. I was certainly behind this thing, knew what it was, and supported it 100%. We tried to get as many people, the faculty and administrators in engineering technology, involved as we possibly could.

O'Hair: So you did provide a lot of encouragement for people like Gil Rainey and Walt Thomas?

McNelly: Yes, and for their faculty. The issues that were discussed in those days were rather key issues that are still haunting us. I keep thinking that every time I look at what people are discussing I get the feeling of deja vu, because we've been through this before and we thought we settled the matter. But we haven't. We always had a small group of people who felt that the real engineering was engineering technology. Well, I have a more operational definition of this: the real engineering is whatever engineering wants to say they are doing now. We are not a stepson or stepchild or less than engineering; we're different than engineering and we support engineering.

There are a lot of similarities to engineering. We're handmaidens to the engineer; we ought to be good support for the engineer. We shouldn't say that we are doing engineering work. I say that today's engineering is tomorrow's technology. I think there is enough work for all of them to do without stepping on one another's toes. We have got to be very, very careful that we aren't really doing exactly what the engineers are doing. If we are, then we ought to be accredited as an engineering program, not as an engineering technology program. The accreditation body (ABET) has a very good grip on the difference between engineering and engineering technology.

I think the thing one must look at is not only the quality for a modern engineering technology program, both in the associate and bachelor's area, but also look to see that the engineering technology people aren't getting into areas that really belong to engineering. We've got to be very careful that we don't. Engineering technology shouldn't be so far ahead in the mathematics and science area that it really could be better defined as engineering. Purdue has a strong engineering program and a strong engineering technology program. I think that has helped both. I think that the strong engineering program would quickly blow the whistle if they think that you're stepping on their toes or on their coattails. I think that's healthy.

The problems in my mind come from institutions that are not really close to engineering. There is nothing to inhibit faculty from really teaching engineering, and it seems like in an educational institution everybody seems to want to be more theoretical. This seems to be the tendency, to leave the applied to somebody else and go more toward the theoretical. It's a normal tendency and one has to simply fight it in engineering-related technology. It's always exciting to be more theoretical and faculty love it. Maybe that's not where the needs are and we ought to be more need-oriented. What are the needs of industry? These are changing, and as they change we ought to change in engineering technology.

O'Hair: I think you have a lot of people [who] would support that. Let me ask you another question. This can be based on your own experience or just your knowledge of what has happened. What in your opinion were the most significant issues and activities in engineering technology that ASEE dealt with in support of entities within ASEE?

McNelly: I think ASEE as a whole has been very understanding and patient. There will always be a small group of engineering technology people who want to be called engineers. There will always be a small group of engineering people who want to kill engineering technology. I think the overall group wisdom of ASEE has simply said, in as quiet a manner as possible, that there is room for both and we need both.

If you let a small minority rule this thing, it will be the end of either engineering or engineering technology. Because of a long-time dominance of engineering, I would perceive that engineering technology would go if we get into that kind of warfare. I think that would be a real tragedy because we would simply have to recreate it as time goes on. There is a spectrum of need and [a] spectrum of talents, and this will always be.

As a psychologist, I can simply tell you that life is not an off/on switch; it's a rheostat, and we've got to recognize that it isn't either/or, but it's many, many shades of gray in between. I think ASEE has shown great leadership in letting this be a family and not trying to let one kill the other. They [have] done a nice job coordinating this, and I think that ABET has done a good job of making certain that the quality levels of both programs are what they ought to be.

O'Hair: One of the entities within ASEE is the Engineering Technology Council, which used to be ETCC, and prior to that perhaps something else. This is the institutional membership. In recent years, this group is trying to be the voice, if you will, for the engineering technology community at the federal government level and perhaps representing engineering technology to corporate America. Would you comment on how this has evolved?

McNelly: I think we've got to understand that this is like shaving. I shaved this morning but it doesn't solve the problem forever. I am going to be faced with the same problem tomorrow, and I think you have the same problem in counseling, but you have to continue to do this. And I think you have to continually educate both industry and the government, particularly Congress, about the differences between engineering and engineering technology.

One of the real problems that I faced when I came into this field was getting Congress to understand the differences. A lot of legislation was passed, and we were left out of some of these bills where we could have benefitted. The Vocational Education Act was a perfect example; it let us become eligible.

The guy that was most influential in this was John Brademas, who was a representative from the northern Indiana district for years and years in Congress. John was an exceptional person who had come from the academic; in fact, he had a doctor's degree from Oxford. He was chairman, I think, of the Educational Committee in Congress. When John no longer had that job, he became president of New York City University and recently retired. He was an exceptional person. Here was a guy who had a Ph.D., who had been a professor before he went into Congress, and knew education like no one else. He is the one, the key person, [who] educated Congress as to the difference between engineering and engineering technology and other technologies. Someday, it seems to me, that we ought to recognize this and give John a special award.

O'Hair: So in your opinion, you think that if we're dealing with the history of engineering technology in general, someone with ASEE should contact Dr. Brademas, for his input.

McNelly: He would be the best person in the world for the people in engineering technology to have come and address them. He's fabulous and I think he might have time now. I think he's president emeritus at NYC.

O'Hair: That's very interesting.

McNelly: I've met John a number of times in Washington, and I came away feeling that here was a guy with great insight into the differences between the various types of technicians.

O'Hair: Let me ask you this, Dean McNelly: you've been observing this evolution for at least, if my calculation is right, 30 years.

McNelly: 30 years.

O'Hair: OK. What progress do you feel engineering technology education has made in the last 30 years, and what are some of those milestones you've observed?

McNelly: There have been so many advancements. I think we're a microcosm at Purdue to what is happening all over the country. The programs have become larger; the public has become more knowledgeable about what they are, and what the opportunities are. Industry has become more knowledgeable of how to better utilize the intermediate technical personnel other than the engineer.

I think that our faculty have improved fantastically. The faculty are able and ambitious. There is a free flow of retired engineers from industry coming into our faculty. This is a very, very healthy thing. I think the fact that we believe in industrial advisory committees is also healthy. While this may seem like a simplistic solution to a problem, it's amazing how many other countries haven't come to it. But it guarantees that we will have a bridge with the user and I think that's important. This has become better as the years have gone on.

Certainly the facilities that we have today are infinitely better than they were 30 years ago, all over the country. We can be very, very proud of our facilities. I think our students are as good or better than they have ever been. There's no problem that we have that would be made worse by having more money. But on the other hand, because of the nature of engineering technology and its relationship with industry, we've had marvelous support, financial and equipment-wise, from industry. This is again something that didn't exist a number of years ago. I think those are some of the major improvements that I've seen. I think it's been onward and upward. Granted there will be a plateau or two, mostly centered around the economy. I see nothing but a greater future.

Of course, the whole problem of whether we should have an advanced degree in engineering technology has haunted the group for the last decade. I hardly want to comment on it. I think it's probably okay as long as we don't lose what I consider our focus, that is on the two- and four-year programs in engineering technology. If we do this, we've got to be very, very cautious about it, and we ought to study what has happened in other fields in terms of emphasis as they evolved into graduate programs. Did they throw the baby out with the bath? If they have, then maybe [we'd] better not throw the bath water out. What we are doing is perhaps modest and limited. Maybe this is our lot in life. If we clearly see a need and our engineering colleagues also perceive a need for us to do this, I think we could cautiously do it. I certainly wouldn't leave engineering out.

O'Hair: **OK. We have already discussed a little bit into the future, but let me go ahead and ask this question anyway: what in engineering technology development do you anticipate to occur over the years?**

McNelly: First of all, Mike, there will always be a constant problem of keeping our laboratories and our faculty updated. This is never going to go away, and, if you don't watch yourself, you are teaching history. We'll leave that up to the liberal studies people. If you have old-fashioned equipment in your laboratory and old-fashioned people teaching in a laboratory, pretty shortly you'll be teaching as it used to be. One has to continually keep faculty current and laboratory equipment current. The two go hand-in-hand, and it takes a lot of creative effort on the part of leadership to make certain that this is accomplished.

As far as what else to look at, I mentioned the advanced degree possibility. These are things that the leadership of engineering technology ought to be looking at, and I know they are, but don't leave out your engineering colleagues. It's so simple to ignore them because you're now fairly big, but you just can't do that. I feel the better programs involve engineering leaders.

We may well have a world leadership role as we go into this next century. We certainly will be involved with engineering technology-related types of people from various countries. Whether we will in the Far East, I don't know, but it seems to me that we are one world now and we've got to all work together and there ought to be a world engineering education kind of activity. I think that ASEE is well aware of this, and I think that it is the kind of thing that we have to address because we're all going to be in this together. We can't just say that spaceship earth is only the United States. It's the whole world, and we've got to help everybody wherever we can.

Recognizing that technology is a two-edged sword, you can have no pluses without minuses. Every advance in technology has a minus. Certainly we have to be more and more sensitive to what technology is doing. If technology is nothing more than a means of better ripping off the earth's raw materials, then we have trouble ahead. I think we are going to have to address the broad issues, not only the right technical background but the right background in terms of politics, psychology, and all those good things that the liberal studies people have been shouting at us for years. Congress ought to get with it and learn more about technology. I see here a real challenge for engineering technology educators.

O'Hair: **That's very interesting. Moving toward closure, a couple more questions. In your view, where is engineering technology today compared to the 50's and the 60's? And in particular, I would like for you to address how you think we are viewed within the total academic community, because we're still kind of the new kids on the block. So, if you can put yourself back into the 50's and 60's as this was evolving, compare that period to the current period. What are some of your thoughts?**

McNelly: Well, Mike, you have to understand a place like Purdue University, where the School of Technology is one of 10 academic schools. (I keep losing count because we just made a new School of Education.) You understand when we first started, we weren't eligible for scholarships like the rest of the schools were. We were looked at as something different. Today, there is no difference. We are eligible for every service that the university offers, just like any other school. I think in terms of Purdue's being a microcosm of the rest of the nation, we have integrated ourselves very well in terms of quality of student and guarding carefully the percentage of liberal arts and non-technology courses in our curriculum so that we don't get too narrow. There is always a tendency to say that if some technology is good for us, more is better, and most is best.

We've got to be very careful about having a balanced program. We can't be too narrow. We've got to keep our people broad because we're going to need broad people that have technical knowledge in the future. And we've got to be very, very careful that again we don't narrow our program too much. If we add another course in technology, we ought to drop one. We still covet the courses we have in communications and psychology and some of these other things that we now have in our program. Our students ought to be literate; they ought to be able to read, write, and do the things that other students do. So, while I think that we are unique in our technical background, we've got to be very careful that we don't get too narrow, particularly when you're in a university.

I think that leadership in the 90's now and in the 2000's [must] address the problems of liberal education. If our universities start to have requirements of a liberal arts background, I think we ought to seriously consider doing it. It doesn't hurt a thing. We ought to try to educate people, not just train people. I think we've got to be very, very careful about being too different from the rest of the university.

O'Hair: **OK. The last question is a very open-ended one. We've talked about a number of things, perhaps some things you'd like to add that haven't been discussed, or perhaps you would just like to summarize or hit some of the high points. So, I turn it over to you; it's an open-ended opportunity to summarize and wrap up this discussion.**

McNelly: I've probably said more than I ought to already, Mike, but I'm particularly proud of what people like you at locations like Kokomo have done over the years. Your hard work and concentrated activities paid off. You've changed the lives of thousands of students and these students add up as time goes on. I look at the School of Technology activities around the state, and in all probability we have well over 13,000 students. This is in any given time, and you think of this compounding itself over decades and how many lives have been affected and how much of the economy has been affected by engineering technology and its impact on the people in this state. These are the things that leave you with a pride that is deserved.

We ought to be proud of what we have done as educators in this field of engineering technology. It has changed the country. We're still the greatest country in the world, and I think that all the activity of educators in engineering that make the technology [has] helped. I just say more of the same. We've got to keep doing it. Thank you for interviewing me.

O'Hair: **Well, on behalf of ASEE, I want to thank you for taking the time to sit down and recall some history.**

Winston D. Purvine

Founder
Oregon Institute of Technology

Interviewed by
Lawrence J. Wolf
President
Oregon Institute
of Technology

Winston Purvine received his A.B. from Albany College (1933) and an honorary LL.D. from Lewis and Clark College (1960). In 1936, he began his career as assistant superintendent and teacher at the Vocational Mining School in Grants Pass, Oregon, and in 1938 moved to the Oregon State Department of Vocational Education, where he supervised on-the-job training for more than 8,500 World War II veterans.

In 1946, he moved to Klamath Falls as director of what is now Oregon Institute of Technology, an undergraduate school offering B.S. programs in engineering, health, and business technologies. Under Dr. Purvine's tenure, OIT grew from a modest 100-student vocational program in a former Marine rehabilitation barracks to a

Oral Histories **259**

modern campus, with 2,500 students enrolled in 12 four-year degree programs.

Dr. Purvine is a recipient of the ASEE's McGraw Award and is listed in *Who's Who in America*, *Who's Who in American Education*, *Who's Who in the West*, and *Who's Who in College and University Administration*.

Lawrence Wolf: **Dr. Purvine, I am going to ask you five questions as requested of me by the American Society for Engineering Education. The first is how and when were you involved with engineering technology in ASEE?**

Winston Purvine: Very early in my career as an educator, I became a job analyzer and was appointed to assist the Governor's Office in producing a classification system for state employees in every department except higher education. It was during this work that I ran across an article in about 1940 or '41 in the magazine *Fortune*, which [described] the New York agricultural and technical institutes.

This so aroused my interest that I corresponded with [people in] Albany, and later on with people in Pennsylvania, Indiana, and numerous other places, where there were either universities offering engineering technology education or free-standing institutions. Some were referred to as proprietary because they were usually family-owned and had been in operation for a number of years. Their operation was due to the fact that they had provided quality education and that word-of-mouth advertising kept them going, whereas many hundreds of similar institutions had gone by the board for lack of support.

It was with this search for information that I came across the Wickenden and Spahr Report which dates back to the early 30's and had something to do with an early study commissioned by the Society for the Promotion of Engineering Education, I believe, to help determine whether there was a need for a hierarchy [of] technical personnel to assist the professional engineer.

This led to my developing a speech about the kind of student who was not at all interested in liberal arts programs, nor were they interested in trade or vocational school programs. Yet they appeared to be of college caliber. So I entitled the speech "The Forgotten Generation" and gave it to every service club and PTA and many teachers' institutes throughout the state of Oregon over a period of some five or six years, preceding the availability of the Klamath Falls' Marine

Recuperational Barracks as a location for a school to provide training for the returning veterans of World War II.

With this kind of rambling discussion of what had gone on over a number of years, it should not be very surprising that the first courses in that general area were the object of rapid upgrading. It was possible to move a vocational course in drafting into a course in surveying over a period of four or five years and to get ECPD accreditation in not very many years, as I remember, eight or ten. And, of course, that called for going to the national meeting of ASEE as soon as possible, sometime [in] 1956 or 1957.

At that time, there was strong discussion in the meetings at ASEE concerning the place and the emphasis for calculus in engineering technology programs. Now I think I ought to say for historical accuracy that the programs at that time were referred to as "technical institute-type" of education. The "engineering technology" name did not come along in general usage until some little time later. In this discussion of the place of calculus, one of the Eastern institutes, I suppose I should say it was Wentworth Institute of Boston, had convinced the high schools in its state that they should be careful to advise prospective engineering and technical institute-type education students at the high school level to take at least one year of calculus and from that, he reasoned, that other states should follow the example.

There were several of us on the negative side of the discussion that ensued because we felt that there was no way that this infant area of education could convince the general educators in our states to require calculus at the high school level. The results of this discussion [were] that the following morning two of the senior members of the group approached me to ask if I would care to serve on the Committee of 21, which was then the guiding committee in technical institute education. I agreed, naturally feeling that this was an acceptance that might even be premature but certainly was welcome.

Soon after that, the Siamese twin of ASEE, ECPD, appointed me as chairman of the Pacific Northwest Region, in which the only institution aspiring to engineering technology was the Oregon Technical Institute. Not too surprising, I had no call to carry out an accreditation visit in those first several months. When ASEE and ECPD met at Colorado Springs [in] 1960 or so, the committee on technical institute education of the ECPD central group was directed to overhaul its procedures in accordance with some of the things that ECPD's main board of directors had determined were desirable and to eliminate some things that they had determined were undesirable. So the first really strong attempt at developing criteria to guide visitors in assessing the value of an engineering technology program [was] put down in writing.

Over the next few years, the various problems that arose and the new developments in the field of engineering technology and ever-increasing recognition of the need for larger numbers of graduates brought about, one after another, attempts to upgrade and better guide accreditation visitors. And because there was, in effect, an interlocking directorship between ASEE and ECPD, the discussion meetings in ASEE were frequently on topics that had to do with ECPD. The result of this was that where there were deficiencies, there were always voices raised to make them known. And also, when it appeared that the standard risers were being overzealous, there were people to point out that condition and discuss it in detail.

The engineering profession, independently but following along the lines of the medical profession, was here and there throughout the country. Partially in educational institutions, partially in consulting engineer offices, and partially in the offices of engineering services in large corporations, [they] were excited about the economic benefits of having applied specialists use the directions of the engineers and engineering planners in developing projects. So the day-to-day, hands-on experience of these people would develop a kind of proficiency that relieved the engineer of routine, time-consuming tasks and permit the application of the professional training that he had received to items that were within his capability alone, thereby making the addition of engineering technicians to the staff a strong economic motivation for continuing and developing.

Wolf: **Dr. Purvine, what in your opinion were the most significant issues and activities in engineering technology that ASEE dealt with? Now you've mentioned the issue of calculus. For the record, I want to point out that you said it was 1957, and that was the year Sputnik was launched. And I am sure that had something to do with that discussion occurring at that time.**

Purvine: Well, I would respond by saying that historically wars have generated some things that they seldom have received credit for. In other words, in the city states [of] the Euphrates and Tigris Valleys, it was the desire to be victorious in war between the states that led to the development of brass and bronze, because bronze would pierce a shield that a copper weapon would not.

Skipping a few centuries, we come to World War II and no single conflict has ever produced as many new scientific, engineering, and medical developments as that war did. And it is because of those developments that engineering colleges went through some changes that I thought were typically illustrated by the dean of the College of Engineering at Michigan State University who, in making an annual address at the American Society for Engineering Education, made the statement, "We have literally plowed under acres of laboratories as we have moved our engineering programs into more advanced science and mathematics technologies."

What then did that kind of activity in engineering education and its discussions in ASEE do for engineering technology? Literally, it put the heat on us to develop more and better technicians and eventually technologists that would be capable of doing applied engineering work, while the planning and the theoretical utilization of science and math and computers was being done by the professional engineer.

Wolf: **Were there any issues other than calculus?**

Purvine: Yes, there were always issues. Some of those issues were related to the things that were happening in engineering. There was some trouble in thinking of our membership in the society that related to just how much hands-on experience we should provide. This was a critical factor and there was always the problem that someone was worrying that we might be confused with trade education because we did some things.

It just so happens in our findings, I think, that I can state a general principle. And that was that an engineering technician who was going to be the intermediary between the engineering office and the craftsman [who] produced the products— say in a machine shop — should at least have a working knowledge of what you could do with a certain machine and what kind of a process would be too expensive and awkward to carry out or what kind of a process, when applied, would produce economic benefits to the industrial corporation involved.

So I'll use that only as an example because there are literally dozens of this kind which the discussions revolved about — partially about image, but partially about realism as to where we should put our resources in the acquiring of equipment, in the selection of instructors, and all of those things [that] go into managing an enterprise as complex as engineering technology became. Because, out of the increasing complexity of engineering, engineering technology found new forces and new pressures that required study of this sort of issue and decisions that could generally be effective in the preparation of students who as graduates could meet the needs of the marketplace.

Wolf: What about the issue of liberal arts training, training in English, these sorts of things? Were they issues at that time?

Purvine: They were hot issues! I can even use our own campus as an illustration. I heard many an instructor say the introduction of a second course in technical writing was taking time away from the necessary skill training that the individual should have in order to be a good intermediate supervisor in the engineering team. It was one of those things; and the way that it got settled, industry supplied us with the answers.

In the first place, most institutions operating at the collegiate level for engineering technology had a stream of recruiters coming to visit them, seeking out those individuals that they thought would best fit into their engineering team in their institution. These individuals were always queried by the instructors at the operational level of the institution. They also were scheduled for interviews with deans and presidents. And from them, we got the answers about technical report writing.

And it about comes down to this: you have two engineering technicians working for you, and you send each of them to review a problem that has occurred in a different place. And [if] one comes back and his technical report is accurate, precise, terse, and valuable, then you don't need to go see what he saw. The other technical report comes back so that the engineer has to go out there and review it himself. Who got the next promotion?

Wolf: What progress do you feel that engineering technology has made?

Purvine: Well, it goes on [in] several different areas. So let's talk about the effectiveness of instruction first. The engineering technology instructor basically has been hired from industry. Obviously, you [need] a practical engineer or a practical member of the engineering team in order to teach an engineering technician the practical arts of [the] field. Also, you [have] to remember that they were people who came from industry and perhaps from engineering consulting firms and whose college courses found few, if any, courses in education and the application of teaching methods.

So there was a general trend in requiring either formally or setting up an atmosphere in which people expected that they ought to go to summer schools at the university level and find some of the theories and methods that would be useful in carrying on their new profession of teaching. That happened [in] many places.

Wolf: **Now that's on the instructional plane. What advancements [were there] on some other planes that you observed?**

Purvine: The follow-up of graduates after they were on the job became almost a religion. Placement officers, advisors from the engineering technology schools went out, but also faculty members who were teaching the specialties like engineering technology and electronics or in mechanical would go see the graduate and find out from him those things that [were] most valuable. And likewise, on the other side of the coin, the things that he didn't need, therefore giving us some clues as to what kind of curriculum revision would be in order.

The next step, of course, was his immediate supervisor; usually that was an engineer, but not necessarily. That person had some basic ideas about what could be done to improve the kind of education that this particular graduate had. Let's do the thing that's important. Let's do the thing that's needed, and let's weed out those things that crept in sometimes simply because of somebody's bias.

Wolf: **What about moving from two-year to four-year programs? What were some of the issues surrounding that as you recall?**

Purvine: The first and most immediate issue was that several engineering deans felt that this attempt was the forerunner of real encroachment into professional engineering education. And we had some hot sessions at ASEE over this matter. And I could call names and name places but I shan't because most of those people who held those views modified them as time went along.

Because what happened? Industry, again providing input into the engineering programs at the professional level, was saying, "We need this," and "this" might be much more advanced computer capabilities; or "We need that," which means that we must have some theoretical math at a certain level, much higher than what we have been seeing, etc.

And as these things happened, engineering professional education moved up in technology, creating a vacuum. The only way to fill the vacuum was for the engineering technician to become better educated, [acquire] greater depth, and that meant a move from associate degrees of two years to baccalaureate degrees of four years and absolutely to the master's degree, requiring additional post-graduate education.

Wolf: **The fourth question on the list is this: what in engineering technology's development did you not anticipate to occur over the years? What surprises did you find?**

Purvine: I'm sure sorry about having to give you this answer. In meetings in ASEE and ECPD and in the discussion of what we were getting from industry, consulting engineering offices, and government locations, we began to talk about all of the possibilities. So it developed, as things went on in the Sputnik atmosphere, in the growth of tremendous developments in computers, in the growth of tools that we could use, and I am talking about a small scientific calculator in the hands of a student which cost him $250 in the beginning but can be had nowadays for $30 or $35.

All of these kinds of things gave a nudge that our imagination had better be working, and it was. So that these discussions at ASEE, both in the formal meetings and in what is called the "cloakroom conversations," these things were so discussed and the imaginations were let free, until we were talking about things that probably (some of them) were impractical. But at the same time, the atmosphere in education and in industry that developed along in the 60's, 70's, and the 80's was such that it was a goad to every one of us, to get that mechanism between the ears really wheeling and think of all the possibilities of how we might better serve not only industry but the upcoming student who had the talent to reach into engineering technology — the interest to be in an engineering team position, but perhaps not quite the capability to be the top-level professional engineering student.

Wolf: In your view, where is engineering technology today compared to where it was in the 50's and the 60's?

Purvine: Well, it's about like comparing the fourth grade mathematics or arithmetic book of the 1940 vintage with an introduction to calculus, the way that it is now being used in the associate degree in engineering technology. It's very difficult to describe this mushrooming of possibilities, of capabilities, and of the kind of investment that has been made nationally, both by public agencies and by several private institutions.

It has been interesting to see that, with all of this type of incitement, many of the fine institutions that were proprietary and family-owned, [such as] the Milwaukee School of Engineering, are still turning out a very high level of graduate. Now that's only an example and it may be unfair to others, but I think the profession is so well acquainted with the contributions of Richard Ungrodt, who finally became president of ASEE and along the way supplied much of the spur on the effort of the people in engineering technology, as well as a great deal of interplay with the engineering deans to help create better understanding between the two groups in ASEE.

Wolf: He was very good at that. I'm going to add a question now. Once you described to me your first trip to Klamath Falls and how many days it took you; I think it was from Salem at that time. Maybe I don't have it exactly correct, but you have seen many changes in technology. Do you feel that you had underestimated or overestimated the changes that have occurred during your career?

Purvine: I am sure that I underestimated; I am certain that very few of us in engineering technology could have forecast in 1947 or 1948 that we would be completely convinced that a master's degree in engineering technology is essential to the development of the economic effort of the United States as it struggles against other developing nations around the globe. No, we couldn't possibly have seen that.

But we did see far enough ahead that we usually had at least three steps in advance in mind. And we took our state boards and our faculty to the first of those three steps; and when that became well-established, we already had developed another step, so [we] were still three or four steps ahead. So we moved to the second step, and so on and so on.

The result being that we followed our nose. But at the same time, what I've said earlier about the statements of needs from the recruiters, from engineering offices, from deans of engineering (and they helped us a lot), from private and public engineering enterprise, whether it's U.S. government, state government, or what have you — in all of this development, there was that continual upward pressure that results from new technology. And so therefore, no. We couldn't foresee it all, but we certainly tried to be alert, to keep in touch with what was going on, to find out where we had best next move. And all of that was brought together at these annual and sectional meetings of ASEE.

I might mention, too (because they were very important and I haven't mentioned it before), that the college-industry meetings in February, [with] interplay between knowledgeable persons in industry, engineers, and engineering technology educators, was a great forum in which you learned many things. You got many hints. This has been a bit like a detective story. You get a clue and don't forget that [it] might lead to something; so you search down the line on it, find that it has no foundation, or that it is a real need; and whichever it is, you are then that much better equipped to handle the problems [that are] going to face your educational institution.

Wolf: **Thank you, Dr. Purvine, the first president of Oregon Institute of Technology, for contributing to this oral history.**

Richard J. Ungrodt

Former Vice-President Milwaukee School of Engineering

Interviewed by
Jack Beasley
Administrator
Purdue University
Programs at Anderson

A past president of the American Society for Engineering Education, Richard Ungrodt received his B.S. in electrical engineering from the Milwaukee School of Engineering in 1941. Concurrent with his graduate study at Illinois Institute of Technology, he worked for Allis-Chalmers Manufacturing Company, where he held a number of engineering positions, including special assistant to the superintendent, Manhattan Project.

Six years after he graduated, he returned to MSOE as a faculty member. In over 25 years of service to that institution, he taught classes for more than 100 consecutive quarters and was instrumental in developing two- and four-year programs in fields ranging from diesel power to electronic communications and computers.

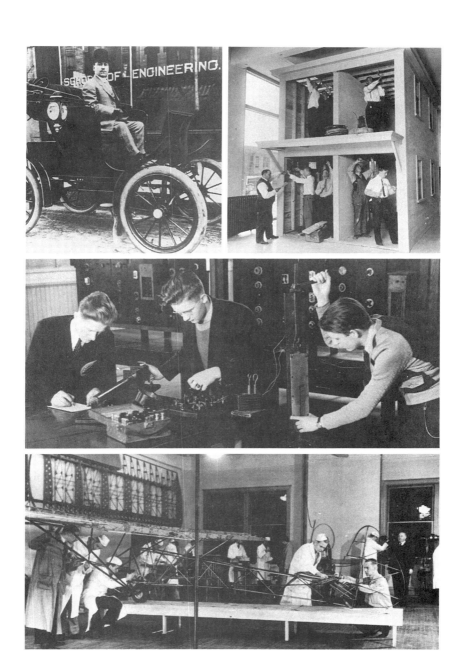

Oral Histories **271**

He has published a number of papers relating to engineering technology education and has served in leadership positions in many engineering professional organizations.

Jack Beasley: I'm here to interview Dick Ungrodt for the purpose of the Centennial Celebration. We are accumulating these interviews with some of the people who have played a big part in engineering technology development over the years. I would like to ask you, Dick, to give just a little background about your career; then we will get into some of the specifics of engineering technology and its development and so forth.

Richard Ungrodt: I appreciate this opportunity to talk with you because there are so many interesting things; you might have to cut me off if I get too verbose.

I graduated from the Milwaukee School of Engineering in 1941. I immediately went to work at the Allis-Chalmers Manufacturing Company in the Research Laboratory, doing high-voltage testing and work on graded insulation for high-voltage transformers and things of that sort. I found it very interesting. There was more routine than I had expected, but it was fascinating. And then when the war came along, I was asked to switch to another facility at Allis-Chalmers.

That's when I got involved in the Manhattan Project. Actually, I got involved in two aspects of the Manhattan Project: one on the electro-magnetic separation process, which involved primarily some of the large magnets in the mass spectrometer that was used for separation; the other aspect was in the gaseous diffusion process in which they were making pumps and products like that to do the same things, but a little differently, in terms of the separation of the isotopes by passing through a centrifugal process and filters. That took me down to Oakridge, and in the interim period I was getting worried that I wasn't doing engineering. I wasn't on a design board; I wasn't making new designs for transformers or motors or things like that.

I think in my neophyte experience as a professional I didn't really understand the problem-solving that an engineer goes through. As a result, I opted for an opportunity to go into the Navy and had a commission waiting for me, which was never picked up because I had been working on the Manhattan Project. I didn't hear about that until later. At any rate, [at] the end of the war my dean called me and said,

"Dick, we've got a lot of GI's coming back. Do you want to come back to teach?"

Meanwhile, I had at Allis-Chalmers been taking graduate studies through IIT at our campus in West Allis. I came back to MSOE in December '46 [or] January '47 to start teaching and taught then until 1988. I went through the routes of all sort of things, started as a teacher which was the lowest level that we had at that time. And then I moved up to department chairman, then to dean and vice-president, and had a fantastic career. I always enjoyed teaching, managed to keep at least one classroom contact every year for 38 continuous years, including the summer terms. It gave me a chance to be close to the students, and I finally ended up with freshman orientation and senior career guidance. I prayed with them as they came in, and I blessed them on the way out. It was a lot of fun.

The one thing unique about MSOE, I think, is its very close relationship with industries and its almost autocratic response to some of the educational units that wanted to come in to accredit at that time. The school was founded in 1903, so when I came aboard it had already been some 40-plus years in existence.

Originally, it had a two-year, pre-college program which was very practical, almost a craftsman type of thing. In 1900 these were kids just coming off the farm, and they wanted to do something technical to get a certificate at that point. But in those two years we gave them sufficient math and sciences that they could basically start an engineering program. Then they had another four quarters which were technician, as we call them today. So that gave them a six-quarter program, basically a two-year program, but they got a certificate at the end of that. Then they could continue on for another eight quarters after that and get a baccalaureate degree in engineering, so it was a plus-plus program. As a result, when those developments came about, those certificates became technician associate degrees.

And then later on, we needed more science in the engineering program, so we had the parallel programs on campus, engineering and engineering technology. One of the key things was that we were very closely related [to] industry. We usually jokingly talked about being owned by industry as a non-profit, non-stock corporation; and, therefore, they had the privilege of supporting us. That was through the very strong advisory committee, and we were one of the very early ones that relied completely on industrial advisory committees.

Beasley: **That is carried through to this day now.**

Ungrodt: Of course. That is one of the strengths of all ET programs today.

Beasley: **That is for sure.**

Ungrodt: ASEE? How did I get involved in ASEE? I joined as a member in 1949. Then in '51 was the first meeting I ever attended, in Michigan. Karl Werwath and Fred VanZeeland, the dean at that time, and one of our industry advisory committee chairmen, Erhart Koerper, consulting engineer, went with us. And that was my first indoctrination.

From then on, I attended as often as I could. Historically, I became an honorary member in '80. In '83, I became a Fellow during that initial period.

Beasley: **What in your opinion were the most significant issues and activities in engineering technology that ASEE dealt with?**

Ungrodt: As we look back on it, there wasn't too much support by ASEE at first. But it was there, and they did pick it up, and they did allow the first few groups to meet together at the annual meetings. I don't even recall when that first meeting was. I was not involved, but there were a number of developments in that area.

Let me just pick up your concern, though. How did we get ASEE to deal with ET? As it matured, it became very viable in the ASEE organization. Then the continuing studies of engineering education recognized engineering technology as part of this. And so more and more as these went on there was reference in the succeeding studies [to] engineering technology as part of that spectrum. Eventually, we had our own studies separate from engineering. That's another whole story.

One of the very interesting things that I see, though, that we did was, through ASEE studies, help develop the guidelines and the criteria for accreditation by ECPD.

Let me just run through the accreditation timeline that I jotted down. I had given a talk back in '87, and that's in the archives at Wentworth, but let me just pick up dates because that's interesting. The first critical one was the Charles Mann Report, and this was a study of engineering education in 1918 for the organization that started in 1903. This is just 15 years, and it shows again that engineering is the most studied and evaluated educational profession that there is.

In 1923, a group met in connection with a meeting of ASEE, technical institutes and technicians in Rochester, New York. This is a group of industry people and education people, and as far as I know, [it is] the first time we have the formal identification of the industry relationship with this group.

Then in 1931 was the Wickenden and Spahr Report for the Society of Promotion of Engineering Education. Oftentimes we heard people joshing about the "Society for Prevention of Engineering Education." But this was a study that had a total of 1,320 pages, and Wickenden did a real good job of studying the whole spectrum. And he included 283 pages devoted to technical institute education, and it was called "technical institute-type" education at that time. He actually identified nine specific functions and activities related to that. That was the beginning.

Well then, in 1941 the Technical Institute Division was established within SPEE, what eventually became ASEE. Again in August of '41, McGraw-Hill, recognizing the importance of this area, published *Technical Education News*, and then they followed that shortly thereafter by the establishment of the McGraw Award for engineering technology education.

Now parallel to this, following those war years, the National Council of Technical Schools, a group of private individual schools that couldn't be part of the regional accrediting groups [and] were therefore barred from government support, needed some recognition, some validity for developing financial support for their programs and for the GI's. They developed educational standards plus ethical standards.

Later on a few public schools got involved, Penn State in particular, because they saw the importance of it. In April of '45, National Council of Technical Schools was the first to establish procedures for accreditation, and they had the first annual conference. At that time they established standards for accreditation, guidance, literature, and they established a representative in Washington, D.C., so they would have a voice.

In 1932, ECPD was established, and it took them to about '40 to '45 before the technical people got involved. Now it's interesting to remember that ECPD was started by the professional societies, and their concern was [to have] a uniform process of evaluation of graduates, [so] that they could determine whether or not they could be acceptable for membership. That's where that accreditation started. Obviously, there was no parallel in the engineering technology. Originally, most of that came about from the chem engineers who were the leaders in this whole area.

By 1940, the TI Division requested the accreditation of engineering technology or technical institute-type programs. In 1945, the first procedures were established, and the first programs were accredited in '46. This then started to be a little group, you might say, that was rearing its ugly head within the engineering spectrum. In '68 and '69, ASEE had a major study on characteristics of excellence; then we moved toward the two-plus-two type programs, but we didn't have differential criteria. That didn't come until '71 [or] '72.

Some of the interesting strengths of the ET movement, which began in ECPD, came about during my activities and a lot of us were involved in it. I just happened to have a hand in a couple of them because of our close industry relationship, and [in] some of the prime institutions their guidance people, their recruiting people, got a little far out in what they were doing. And so finally, in '62, [we] required every accredited program to submit their catalogs for review of their publicity policy statements. That started some concerns; it also gave us a chance to show engineering that there was a proper interface between engineering and engineering technology because monitoring was very important.

Now up until this time, we had not had any professional society involvement in ET. They were considering us as a little sister down there. ASEE had taken the leadership in supporting this [by] providing the accrediting members, members who were a part of ECPD, and there was no real demand that they had to represent any particular professional society. As a result of that strong activity and the strong programming activity by ET in ASEE, professional societies started to take a very careful look at that whole thing, and they developed some interest in it. It was ASEE that provided the leadership for accreditation directions and the accreditation members to guide it, as well as the guidelines.

In 1970, the liaison came up with the relationships with professional societies, and both ET and EETA then started to come together a little more in terms of similar criteria. In 1972, the ET commission said, "OK. Professional societies can provide members for visitation and evaluation provided they show what was called 'demonstrated interest' in ET." That gave us a chance to go back to the professional society and say, "Look, ASEE has provided these people all along. Some of them have been members of your society and sometimes not."

But "demonstrated interest" means that they had to have some means of recognizing the technology graduate in their society so they could have some membership base, because that was the direction really for accreditation, through the membership. We said, "You probably really should have a student branch support in technical institute or ET programs" so that professional societies were represented on the campuses. That was the general nature of that development.

From '72 to '77, this developed within ECPD and with the pressures of ASEET groups or TI divisions. And then by '77, we had the first curricula listed by participating societies. Up until then, it was just a general listing by societies and not the separate listing by societies. So again, engineering and engineering technology were coming closer together. Then by 1980 we had ad hoc visitors supplied by societies, a good move. ABET was developed out of ECPD, and they were just concerned with accreditation, so there was technology accreditation and engineering accreditation.

It is interesting, too, how the coordination came together. Actually, in the ECPD we had fairly good respect from and understanding of most of those people. There was one time, however, in the Denver meeting that one of the deans from the West said, "Let's throw the rascals out." That was rather an interesting thing because then the other deans began to identify where they [stood]. It was a very healthy exercise for us because we had to take that criticism and identify it.

Now I was in a good position because we had both engineering and engineering technology on the campus, and we lived together. For some time, a lot of people looked to me as a spokesman, although I don't think that I was a good spokesman. At that time we were not a part of the North Central accrediting group, because we had a group of rugged individualists on our Board of Regents who did not want the educational group telling us how to run our programs. But later on, we certainly came to full realization on that.

Beasley: **What progress do you feel that engineering technology has made over the years? That's a wide open question and you have already covered part of it, but can you focus on that aspect?**

Ungrodt: Let me think. A couple of areas: one, if you're looking at recognition for ET, so what is this....

Beasley: **That's an ongoing thing.**

Ungrodt: That's an ongoing thing because of the very, very close interface and relationship with engineering. And not only in the educational programs, but particularly in the output side, in the graduates. Graduates get together out there and they interface. Certain graduates have more opportunity to become involved in the engineering type, and others get more involved in the practical.

Beasley: **It has been my experience that the engineering technology graduates become engineers when they get out in the field many times, or the other way around: engineers become technologists.**

Ungrodt: As we define it, yes, I think so. And one of the reasons for that is, I think, that engineering has never structured its educational program like the other professions. You look at any of the other major professions and the master's degree is a minimum entrance requirement. We don't require that in engineering. As a result, I think that is one of the things that I was disappointed in seeing, that engineering did not go that route because that would have given us a tremendous opportunity for a really truly concentric-type program. A lot of our engineering graduates go out and become technicians, and they're dissatisfied. They think they should be doing all that fancy design. I felt the same way when I graduated.

Beasley: **They want to reinvent the wheel.**

Ungrodt: That's right, but engineering is problem-solving. It becomes very practical, and it's very natural that when they solve problems of one discipline that they can solve problems in another discipline. Then they become managers, and they move up the line because they are good organizers and good evaluators, and you know the whole story. But that is one of the problems that just hasn't really been settled because engineering still has its foot in the practical, as well as in the theoretical.

Beasley: Being new in this field, I have a hard time separating engineering from engineering technology. All of us — well, 99% of us — are engineers by training that are involved in engineering technology. We are, I guess, converted engineers. I have experience over the years of having a lot of engineers work for me. What I really needed was engineering technologists, but I didn't know there were any. And there weren't any in my early years.

Ungrodt: I really see that as sort of an artificial designation, engineering technologists. We were involved in the Mideast with a couple of area schools that had engineering technology plus engineering, but the University of Petroleum and Minerals in Saudi Arabia had a group of us over there back in '72. They [had] two programs: one was engineering science and the other was applied engineering. You know what the applied engineering was. They couldn't use it as engineering technician as that terminology translated in Arabic to "engineer's little helper." It's very demeaning to that Mideast attitude.

The other aspect that didn't develop as I hoped it would was the general recognition of the nature of the engineering technician so that he could become registered. The four-year graduate has more than enough to become registered as a professional engineer. And in many places there is a real artificial restriction. Somebody with a B.S. in physics can get registered, but somebody with a B.S. in engineering technology cannot.

Beasley: They either restrict them in the time that they have to serve before they can go up for registration or [in] many ways.

Ungrodt: That's right.

Beasley: My son is an EET graduate and he passed his engineer-in-training the first time through. He is going up for his professional this fall.

Ungrodt: This is interesting. In Wisconsin, they can take the two exams on two consecutive days, so we have always encouraged our graduates to go up. I don't know what the recent record is because it's been [several] years now since I have been directly involved in following them. I used to follow them pretty regularly. The ET graduates did much better on the practical exam, the second exam, but didn't do so well on the fundamental exam, which is natural.

Beasley: The next question is, what in engineering technology's development, and you've touched a little bit on this, did you anticipate would occur over the years? What surprises have you had?

Ungrodt: I've touched on everything but the two surprises. Probably based on the attitude that we had at our school — that this is the way to develop the two-plus-two type for the American standard pattern, expanding scientific and more practical knowledge. I did not expect that engineering would stay this long with the requirement of the baccalaureate for the first professional degree. And that has blocked the development of the bachelor in engineering technology. And if that had come about, fine. The other is recognition by the state in terms of the work they are doing on the certification and registration.

Beasley: Is this true throughout? Every state has its own registration?

Ungrodt: Right.

Beasley: And different rules in almost every state?

Ungrodt: From what I have seen, yes. In one state in the South, I was asked to come in and evaluate the program for credit for parallel registration. I had a great result in terms of putting the thing together and showing the parallels and the similarities as well as the unique differences. Working with the chairman of the board of registration, who was the dean of engineering of that institution of that state, was a dead blank. Absolutely not. It couldn't really be justified. He justified it primarily in terms of their passing rate. Yet the school had a lot of foreign students because of the strong program. Foreign students didn't care about registration so they would go up sit in their room, take out the paper, and walk off. So the statistics were not good.

Beasley: This is what I've been through. The hardest thing to do is to educate people as to what engineering technology is. I remember when I interviewed for my first teaching job with Mike O'Hair, I knew very little about technology; and they specifically in the interview asked me some pointed questions just to see if I was prejudiced about what they described as a technologist versus an engineer. Luckily, I had enough experience in industry to know that there was a very big need for technology people so I was just all for it. So maybe that is why I got my job.

Ungrodt: I'm sure. You know that parallels the ASEE structure to a great extent. When the Relations with Industry Division was established, the different banquets would be scheduled at the end of the meetings and the TI banquet was usually on Tuesday. And then the RWI was the next night, and all of our TI people were over at the RWI banquet. And then the following night on Thursday we had the annual banquet. Now they have compressed that a little bit so that it's hard to get to all of those. The result of those very close contacts when the College Industry Education Conference was structured, one of the strong divisions, one of the strong groups, was the engineering technology TI Division.

And when you look at the ASEE annual programming, you had the different divisions: engineering, electrical engineering division, civil engineering, mechanical engineering, and so forth. They would have nothing to do as divisions with the technology programs; those were structured by the deans or the department chairmen, and they felt that engineering technology had no place there. As a result, the ASEE TI Division, or ET Division, started to structure and expand programs for improvement of the curriculum in electrical, mechanical, and civil—these were parallel programs.

Pretty soon when the ASEE Board of Directors sat down to look at the annual meeting, they were getting a tremendous number of technical institute-type or ET programs within their system very naturally. That's the one area that I am a little discouraged about seeing; the other divisions, the engineering divisions, have not really worked together with ET divisions. It would have been a good thing, because together we can help to structure our programs to interface properly so that some of the common areas can be followed, as well as some of those that are separate, and we understand where each one stands.

Beasley: If you don't mind, I have a question that is not on the list here. In the last 10-15 years, have you found industry in general beginning to accept engineering technology graduates as equals to do engineering or at least accept them? It has been my experience, at least 10-15 years ago, some companies did not accept them at all, would not hire them. Now I am seeing them being hired more and more; in some cases, they are being hired more from technology than engineering. What has been your observation?

Ungrodt: I have not been in close contact for the last four years, but let me comment that it varies by company; and it depends sometimes on the school that the hiring man came from because he understands that program. He says, "This is what we need." When you come to our program, they will hire engineering technology graduates and engineering graduates, but not necessarily for the same area. They look at the functions of engineering a little differently than we do in education, because in education we ignore the functions to a great extent.

Although we do put some management in there, we don't really focus on the need for the graduate to understand that when he gets out in industry he's not going to be sitting there with a calculator and a computer doing design all the time. He will be solving problems of a wide variety, might get involved in manufacturing, might get involved in the working of the labor groups, customer relationships.

Beasley: Be quickly into supervision.

Ungrodt: That's right.

Beasley: Getting back to the list here, in your view, where is engineering technology today compared to the 50's and 60's? Again, you touched much of that, but....

Ungrodt: I think it's come to be recognized in many circles as a legitimate, identifiable area of engineering. The fact that we have to hang engineering technology down there gives us some troubles as compared to functioning as an engineer in industry. I am a little concerned that we just have not really broken that completely.

Where I think we are farther ahead today: we have our general engineering education that gives it a particular aura that people can look to and say, "Yes, I understand that's different than the general that you find in electrical engineering or mechanical engineering." Yet they see the interface; it's always there.

I would hope, looking to the future, that engineering would finally do what it should. That's one thing that I projected years and years ago; that by the time I was through with the profession, the master's would be the minimum professional degree requirement for engineering. But I think the high demand and need for the graduates and the industry's willingness to take them at that level and need them has just said, "Why should the school require that and the industry doesn't require it?" It's primarily, I think, when you get with registration and civil service, some of the things of this nature, registration was another question involved.

Beasley: **We have covered the basic questions. I think you have a few comments on several various subjects that you would like to cover. We would appreciate hearing those.**

Ungrodt: OK, thanks. One is the contacts in Washington, D.C. Very early on after the war, the National Council of Technical Schools had an office related to Washington, D.C. to help get funding. That disappeared as ASEE took over in that area. But more recently, when ECPD brought in engineering technology and wanted to move to the baccalaureate level, we had to justify it with the National Commission on Accreditation. ECPD had to justify its existence in the ET area.

Now the engineering area was sort of grandfathered in. Then there was the Federation of Regional Accrediting Commissions on Higher Education, and they said there must be some consensus because everyone was trying to come along with accrediting programs. And, as the result of that, the National Commission on Accreditation was basically the accrediting agency to accredit the accrediting agencies.

As the result of that (I happened to be chairman of the ET committee of the ECPD at the time, and David Reyes-Guerra represented the EETA), we went into Washington and presented our case to the NCA, National Commission on Accrediting: that ECPD should be authorized under NCA to do the accrediting in the technology area. That was a day-long session; it was very interesting, and we were accepted without question.

That was our first experience in Washington and that led ECPD in the ET area to start to bring their programs a little more closely together in the terms of the format that engineering had. It also took engineering and brought it closer with the industrial advisory committees and industry representatives on the visitation team. In fact, we had established in ET not only the publicity policy statement review but also the requirement that there would be one industry representative as the evaluator so that you would not get just the dean's academic view. The other one that was different from EETA was that we said in the second or following visit there must be one who attended the previous visit, so that you would have continuity. The EETA said, "No, you don't want anybody coming back who was there before," but that worked out pretty well.

Now the second area of activity in Washington was working with civil service. Again, because of the difficulty of becoming registered as a PE (many civil service positions required that registration), a group of us went to Washington and met with the civil service agency and went through the same process, spent at least a half a day discussing with them the differences and the similarities: where our engineers were, where technology graduates were in terms of being accepted and passing examinations. We finally got a tentative statement that looked very positive. When it came out officially, it said, "Yes." As soon as it came out, some group in Washington got on an inside track and the civil service group retracted that. So we lost that battle.

Another area that parallels that was the relationship of engineering technology with the National Society of Professional Engineers. NSPE has members who, as a result of being certified in their state as professional engineers, are registered. The engineering technicians had no place for that, and early on, many of the industry people in NSPE said, "Look, if we don't bring these people into the profession, they are going to end up in the unions. Let us help found an organization for technology graduates," technicians at that time.

So they had then the Junior Engineering Technology Society, JETS. They worked later on guidance, but then they had certified engineering technicians. Now this was done voluntarily by an organization through the NSPE, as a sub-committee of NSPE. That gave them a parallel type of recognition that the NSPE members had by reason of registration.

That's where the whole thing finally developed. The one person that I mention to you that you probably should contact is Curly Foster. He has had a lot to do with the NSPE and the activity in that time frame. He presented a couple of papers in that area. As close as we have come, that is one area that we haven't really jumped in.

Beasley: Any other subjects?

Ungrodt: Well, you know when you asked me about this and with the failing memory you have at age 73, I went back and looked at some of the developments in ASEE from the early days that I got involved in. I started teaching in '47, and ASEE's first report that I ever saw was a report on the 25-year anniversary of engineering technology education at the 1948 meeting. So somebody went back and said in 1925 [that] we had these programs. Of course, it was already recognized in the Mann Report in 1918, but it wasn't quite as structured [then].

The interesting thing is Dean Boelter, a very strong engineering dean, was the speaker at that 25th anniversary meeting of what was called technical terminal education. That was a two-year program. Well, that helped us pick up the timeline and keep it short. In 1950, we had the first ASEE McGraw Award, and it went to Harry P. Hammond at Penn State.

Then another interesting point came along in 1954. ECPD could list a separate annual report for the engineering technology programs. Previously, they had sort of been a sub-committee, and they were still a sub-committee at that time of the EETA Committee and part of that sub-committee report. Apparently, some of the engineering deans didn't like that engineering technology report in this Engineers' Council for Professional Development annual report. That was then set up as a separate report, and there were 28 separate institutions and programs. [The] next year it went back into the regular annual report.

Then in '64-'65, engineering technology became a separate standing committee. Previously, it had been a sub-committee of the EETA in ECPD. That's when we met the National Commission on Accrediting requirements, because as a separate committee we were then coming away from the umbrella of the EETA and had to have our own criteria. [The year] 1955 has always been a landmark position in ET. You would recognize the ASEE Grinter Report, which was '53-'55. It's rather interesting. Let me just read a couple of real quickies just for the record.

Beasley: Sure.

Ungrodt: Because the first report is real hard to get. I had a mimeograph copy from one of the original members many, many years ago, and I've got that almost bound in gold leaf or something so it doesn't get lost. The actual wording of that preliminary report is really interesting as we look at developments which have occurred since that time in the baccalaureate and technology programs.

The report states, under definitions of professional-general and professional-scientific curriculum:

> *There seems to be no major disagreement that an engineer cannot be trained to make effective use of modern knowledge of engineering science and creative design within a four-year undergraduate program.*

The engineer cannot be trained within four years for that.

> *It is even more improbable that effective contributors to research in the engineering sciences whose development is now an accepted responsibil-ity of the engineering profession can be trained in four years. The only questions that remain are how the professional scientific educational process is to be conducted in what undergraduate background is consid-ered to be necessary preparation for extension of the student's education for scientific or analytical phase of engineering. It seems more probable that four-year training may be sufficient college preparation for many students with general professional objectives. To accomplish these different objectives engineering education in the future should be more functional than in the past. The committee recognizes that at present there's substantially no difference between curricula of the same designation. If engineering is education to serve two quite different demands — 1. supply engineering personnel for production, construction operations, sales, installation of engineering equipment, etc. 2. supply professional engineers and engineering scientists able to interpret and use for design purpose the information being provided by research in the engineering sciences, and also to advance the fields of engineering science there will be need for greater functional divergence in engineering education. The first four years of the undergraduate program can hardly be identical any longer for these two types of engineering education which in this report will be called professional-general and professional-scientific.*

Now that was really far-reaching. The conclusions of that preliminary report were five in number and indicated the need for higher accreditation standards for engineering curriculum. The fourth recommendation seems most critical for setting the stage for the growth of the baccalaureate programs in engineering technology and reads as follows:

> *The functional divergence so evident in engineering activities which range from research to management has led to the committee's recommendation that accreditation be based upon either of two designed defined functions in engineering education — professional-general and professional-scientific education.*

Well, what happened is that this report went back to the engineering deans and as they reviewed it they could not see, during the war years with all that emphasis on master's research, the two programs standing side-by-side. And as a result, it was by a wide margin turned aside. They never did move to that master's program. And Grinter really had his finger on this thing, saying, "This is the future." And that, I think, is what happened. Then they called it engineering science and forgot the engineering practical. That's when the baccalaureate programs, IMET, moved in.

It's interesting that there was a misinterpretation of his reports because many of the colleges as a result removed practical laboratories and [went] with more scientific, more engineering sciences. As a result, industry said, "That's great; we need this man with this great scientific knowledge. We will train him to be a practical person." But in the meantime, the practical areas for a lot of the small companies [were] not available. And that is when the four-year technology programs really started moving.

Beasley: I started in electrical engineering in '58, so I started right at the ground floor there at Purdue when they switched over.

Ungrodt: That's interesting.

Beasley: So I have a feel for what you are talking about there. My cousin graduated in '49 from Purdue and what he took and what I took were completely different. He had courses like surveying, drawing, and things like that. It was just completely different from what I took.

Ungrodt: There was a lot of routine design, and you had to have all the functions and understanding of it.

Beasley: **I had very few lab courses.**

Ungrodt: Grinter didn't really mean that. They misunderstood the report findings. They didn't look at the preliminary report and say, "What is it based on?"

Beasley: **I think it was an over-reaction maybe.**

Ungrodt: I think it was.

Beasley: **But we should be glad for it.**

Ungrodt: Yes, it identified technology, and it has identified technology as a very special type program that is needed by industry. Then in '57 came Sputnik and further emphasis on the scientifics, and they never had a chance to come back to it. That's when there was a real emphasis on the four-year. [In] '65 to '66, Dean Lohmann was involved in identifying the characteristics of the four-year program. Prior to that time, ABET or ECPD had only the two-year programs; the four-year programs that were accredited using the same two-year criteria for the four years. Those differential criteria were established in '71 to '72 by the accrediting group. The beginnings, in '66 to '67, were based on an ASEE study that preceded that.

Looking back historically, there were two separate commissions: the engineering technology commission and the engineering commission. There was also the accreditation coordination commission and that group worked together so ET and engineering procedures would not be too divergent, but they would not get in each other's camps. That diffused a lot of the pressure from many of the engineering deans who didn't want anything to do with technology. As a result of that, some of the schools picked up engineering technology. There were a few others that dropped their engineering technology and kept their engineering.

Beasley: **That is still going on, isn't it?**

Ungrodt: Oh yes, it is. I was privileged to be chairman of an accreditation group to Cal Poly at San Luis Obispo, where General Higdon lived at the time, and Gene Nordby was chairman of the EETA Committee. So the two of us went out together and put a common team together of engineering and engineering technology. I ended up as general chairman because there were more technology curricula present than engineering programs. So we have had a chance to work together seriously.

Beasley: **Very interesting. So are there any other subjects you would like to cover or summarize?**

Ungrodt: It's an exceedingly dynamic situation. I think the dynamics we see today are different than they were in the past. We had pressures, external pressures after the war, for education and for an education to meet the needs of industry. Then we had concerns after the war for the Sputnik question. We had been on the upgrade economically at that time and there was a need for all of these graduates. When the need got particularly difficult to satisfy, industry would take whomever they could with the right background. That's when the engineering technology graduates started doing engineering as well as they could, much better than some of the engineers who had not had enough of the practical courses.

More recently, if the economic times turn down, we'll probably find some people who will say, "Well, he's an engineering technology graduate; we won't take him. We'll choose the engineer." Only because of the aura of sophistication [that] comes around on that. Those companies that have had better success with engineering technology graduates will probably opt for the ET graduate.

Beasley: **But in light of the necessity now for this country to copy Japan, let's say, being more practical, not trying to invent things, but to take things and make them practical, wouldn't an engineering technology person fit into that mold?**

Ungrodt: Absolutely. I would like to see a basic undergraduate degree that's practical, given enough of the sciences so that they can learn. Then when those people graduate, they have the opportunity to go to work; they would have an opportunity while working to continue their education in a functional area rather than in a discipline area. As they do certain functions in an industry, they might try management, production and manufacturing, or whatever it might be. That's when they should come back and do their studying. Those who are more scientifically oriented, let them go on. But we need about, I would say, ten of the four-year ET's to the one engineer. But we should be calling them engineers. They disappear into the woodwork of industry and they show up as engineers.

Beasley: Exactly. I have never found a company that has an engineering technology category here as a classification. I suppose there are.

Ungrodt: There are a few that did have.

Beasley: **I know in my son's case he went to work for GTE and he suddenly became an engineer.**

Ungrodt: That's right.

Beasley: **He said nobody ever asked him the difference — you know, he's an engineer.**

Ungrodt: Well, by the very nature of the four-year degree graduate, he doesn't work under the direction of an engineer any longer.

Beasley: **That's true.**

Ungrodt: A two-year tech probably did. But those have even become quite sophisticated.

Beasley: **Well, we could probably go on all day, but I think we have covered a lot of history. I am very glad to get the chance to get to talk with you, Dick. Thank you very much.**

Ungrodt: Thank you for the opportunity.

Eric A. Walker

President Emeritus
Pennsylvania State University

Interviewed by
Joseph DiGregorio
Associate Dean
Pennsylvania State
University

Dr. Walker is president emeritus of the Pennsylvania State University, having served as president from 1956-70. He also retired from the Aluminum Company of America, where he was vice-president for science and technology.

With a Ph.D. in electrical engineering from Harvard (1935), Dr. Walker taught at Tufts University and served as consultant to electric utilities on high-voltage power transmission, a field in which he holds several patents. He has worked for the Department of Defense in several capacities and has been president of ASEE and the Engineers' Joint Council, and chairman of the board of the National Science Foundation and the National Academy of Engineering.

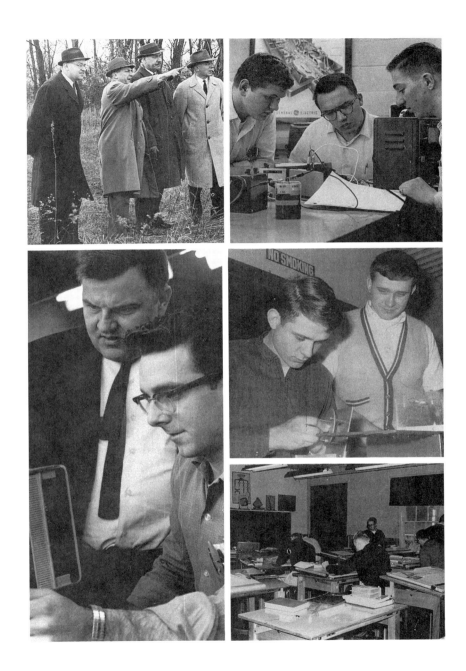

The author of more than 300 publications, Dr. Walker has been a director of 10 industrial corporations and advisor to industry on research and development programs. In addition to ongoing involvement with Penn State, he also consults with other universities on problems of planning, financing, and organizing new ventures, such as outreach programs and affiliation with high-technology parks.

Joseph DiGregorio: **This is Joe DiGregorio, associate dean for Commonwealth campuses at Pennsylvania State University. I am about to interview Dr. Eric A. Walker, president emeritus of Penn State University and dean emeritus of the College of Engineering at Penn State.**

DiGregorio: **Dr. Walker, how and when were you involved with engineering technology in ASEE?**

Eric Walker: Well, I was in engineering technology before I ever became a member of what was then called Society for Promotion of Engineering Education. Because back in 1939, I went to work for the General Electric Company in a division which designed transformers. It immediately became apparent that very little of the engineer's time was needed to design transformers. What you needed were engineering aides who could do arithmetic, select wire sizes, and so on, which a technician could do just as well [as], and probably better than, a full-scale engineer.

Then later on, we got into the real work and we couldn't find engineers. They were all working in industry, and we needed people who could put together circuits, test them, and so on. And I got the idea [that] the best people to do this were radio hams. So I went all around New England hiring people who were members of the AAARRL, the amateur radio league, and we made great use of them. Then we started hiring women and training them for doing this. In fact, we hired probably 50 girls from the Junior League, because these were intelligent people, who could quickly learn how to test writing diagrams, do the blueprinting, and so on and so forth.

So I had a genuine interest in this long before I came to Penn State. But shortly after I came to Penn State, Ken Holderman and I began doing some arithmetic, and this was the time when there was a great shortage of engineers.

And all you had to do is take the number of high school graduates; seven percent of them said they were interested in engineering. He took that seven percent and multiplied by the number that were going to graduate. And immediately you saw there'd be a great shortage, and so we went back to our old idea of getting intelligent young people who were interested in working with their hands to become engineering aides. So Ken and I put on a great campaign. First we had to sell it to the trustees, get them to decide to give a degree of engineering associate.

DiGregorio: **What years are we talking about?**

Walker: This would be 1951 or 1952, somewhere along in there. Well, selling the trustees wasn't very difficult because a good many of them were engineers in managerial positions and they knew they needed engineering aides. The next problem, of course, was to get a curriculum, and it was very difficult to get the faculty interested. They couldn't believe that there was a market here. They thought it was low-level engineering. We tried to tell them it was a different kind of engineering. But finally, we got enough of them interested, and we started to write courses and curricula.

How do you get students? And again the parents and the high school counselors thought this was low-grade engineering. You don't take the best students; you make them into engineers. And again we had to explain that some people want to work with their hands, and we were going to have a curriculum to do that and to teach them the kind of thing that would help them to be engineering assistants. I think Ken and I talked to almost every high school counselor in Pennsylvania. We had dinners for them all over the state, and finally we did get a body of students who would accept this kind of education. Of course, the payoff came when the first class graduated, and two companies took all of the graduates, and those two companies kept on taking the graduates. At times they would pay the graduates of the engineering associate's program more than they would the graduates of the B.S. program.

So we got it sold, and we looked around for other schools [which] were doing the same sort of thing. The only one that I can recall in ASEE that was really sold on the idea was Purdue. So Purdue and Penn State walked in lock-step, and I think we finally got the program to where we wanted it. But even when we talked to people in ASEE, there were many people who were very skeptical.

Though many people would start the program, they'd let it backslide. They'd backslide in two ways: they'd go down the scale in what they taught them, or what they would try to do is bring it up and make it the first two years of the regular baccalaureate program. And neither answer was any good, you see. I think we still suffer from this idea of many people thinking that this is the first two years of an engineering program, and it isn't. It shouldn't be.

DiGregorio: **Dr. Walker, what in your opinion were the most significant issues and activities in engineering technology that ASEE has had to deal with?**

Walker: Well, I think the very deficiencies that I've already mentioned, but ASEE began to hold sessions on engineering technology and many of the lesser-known schools embraced the idea of having those sorts of things. I was always disappointed that schools like MIT, Cal Tech, Stanford, and so forth chose almost to ignore it. Now, it might well be that this is the right thing to do, because if they took it up, people again might think this is a dilution of a four-year curriculum.

And the talk among ASEE, especially at the winter conference (the CIEC), I think people understood what was wanted, because there [were] so many industry people there who would say, "We would hire these people if we [could] get them." And I think a great deal of progress has been made, but it is going to be a continual fight to maintain the specific character of the engineering technology program.

I think that I did not anticipate the fact that even today there are a good many engineers who are working far below their skills. For instance, they get hired by NASA to run the launching program, and that does not require a four-year degree program. A person takes that job; he stays in it one year, [and] he's done everything that has to be done. He starts his second year and he repeats it. He starts his third and repeats it again, and at the end of five years he hasn't advanced his technical skills one bit. And then he wonders why a new company comes in and says, "Your salary is quite out of line with what you're doing. You're fired, and we're going to hire a new graduate to do it."

You ask if ET was considered a separate discipline from engineering or really applied engineering. I think it is both. You don't need to understand triple integrals to be an engineering technician. There's no point in giving you courses in higher math. So, in that respect, it is a separate discipline. But again it is applied engineering.

One of the difficulties with the B.S. program is it produced people who [didn't] know how to do anything with their hands. We've got a high percentage of engineering professors who have never done any engineering. And, if I had my druthers, I would say no one could teach engineering unless he's actually done some engineering. You wouldn't ask a surgeon to teach surgery if he'd never done any surgery. We ask engineers to teach engineering students when they don't know what engineering is. They know research, they know the theory, but they don't know the practice of producing something that people want, that they can afford to buy. They never worry about "do people want it," but "can people afford to buy it."

DiGregorio: **As you were speaking, Dr. Walker, it occurred to me that in its infancy, engineering itself was considered applied science. In the same way, as we look at engineering technology as applied engineering, do you think eventually engineering technology at some distant, future time will be a discipline of its own, just as engineering became a discipline of its own, separate from science?**

Walker: I suspect it will because we've never used the term "applied science" for engineers and yet that's really what they do. The definition of engineering, which I learned when I was in college, was [that] the engineer's job [is] to take a good idea, and using manpower, money, and materials, produce something that people want at a price they can afford to pay. And the emphasis is on production.

Today, we are learning that we are way behind the Japanese in producing things: in taking these good ideas and not researching them, but developing them, designing something, manufacturing something, and selling something. And, of course, most of our engineering professors have never done this. And I would be quite happy if someone started a movement to say that any engineering professor, before he can be called a professor, must have done some applied engineering—producing something that people want.

DiGregorio: **From another point of view, the American system has been criticized because we tend to separate our basic research from our applied research. We put our research organization in city A, if** we're a company, and in city B we put our manufacturing plant, and in city C we might put our sales force, whereas the Japanese put them all under one roof. Is there a danger that if technology becomes a separate discipline that we will fall further behind the Japanese? Don't we need this integration to continue?

Walker: Well, this is true. If you take General Motors, or even General Electric, when I knew them, they had research laboratories with the research side (pure research they called it) and applied research. But then to carry anything from applied research to engineering, you had to go up through a vice-president for research, a vice-president for engineering, back to the engineering. To carry it from development to manufacturing, you had to go the same way around.

And, as a matter of fact, [in] the most successful companies, often small entrepreneurial companies, the man who had the bright idea followed it through research, development, design, going back for more research if necessary, design to a prototype, a prototype to manufacturing, and even to the sales department. The Japanese learned this very early. And some laboratories in the United States learned this. The Naval Research Laboratory, our own Applied Research Laboratory puts it all together and now begins to say the same sort of thing.

Of course, one reason is that we had a National Science Foundation. When Van Bush started to put that through, he wanted it to be called the "Science of Engineering Foundation." But, as you know, Truman vetoed the first bill. The second went through. "Engineering" got dropped off and "science" was the word. And when the National Science Foundation was founded, it had only one engineer on the board, and they forgot engineering. We didn't even have an engineering division for 10 years. And engineering professors decided if you're going to get a grant from the National Science Foundation, it had better be research, and the purer the better. This [leads] off in the wrong direction.

DiGregorio: You talked about yourself and Ken Holderman as being two of the pioneers at Penn State in engineering technology, and you mentioned Purdue University. Can you name some of the other people at Purdue and the other schools you feel were really responsible for the emergence of engineering technology?

Walker: The only name, and I knew a lot of people at Purdue, but the only name I recall offhand is a guy named Gere, and he helped us write the goals reports. And, unfortunately, in the goals reports, we completely neglected engineering technology. I don't know whether we were told by the National Science Foundation to neglect it, but I looked through that book the other day, and there were only two short recommendations about engineering technology, saying we ought to study it and we ought to do something about it. But nobody did.

DiGregorio: Where was Gere located? Where was he?

Walker: He was at Purdue. [On] page four in the "Goals of Engineering Education," there is a footnote which says "the need for supporting personnel, especially programmers, technicians, etc., would be greatly served if a comprehensive national study were undertaken which would involve representation from technical institutes, community colleges, engineering colleges, as well as the professional organizations concerned with technician, engineering, and scientific personnel."

DiGregorio: In what year was that written, the footnote?

Walker: 1965.

DiGregorio: That was right before the baccalaureate degrees in engineering technology became available?

Walker: That's right. [On] page nine there are graphs, you see, of the present supply, and the needs in the next decade, which covers "engineering's opinions of the adequacy of the present personnel and the relative needs of the next decade."

DiGregorio: What were some of the early programs that Penn State offered in engineering technology? I know there was a design and drafting program.

Walker: I think we had design and drafting. We had electrical technology, thinking of people who could repair television sets, who could wire up circuit boards in research laboratories, who could test circuits, and so on. We had one in mechanical engineering, obviously: the same sort of thing, where people were going to be able to run the test cells, test motors and engines, and so forth. And I'm sure we had one in civil engineering, but I can't recall exactly what was in there.

DiGregorio: **Our first graduates were [in] 1955 or so?**

Walker: Probably.

DiGregorio: **When did we begin the bachelor's degree in engineering technology?**

Walker: Much later.

DiGregorio: **Mid-60's, late 60's perhaps?**

Walker: Probably later than that.

DiGregorio: **There is now a push in the technology field toward merging electrical and mechanical once again.**

Walker: I think we should. I think we should.

DiGregorio: **We are thinking about doing that ourselves.**

Walker: Mechanical and electrical can go together, civil engineering and surveying. You know, you don't need a full-scale engineer to do surveying today with all the equipment they have. You need a technician.

DiGregorio: **Dr. Walker, this is ASEE's centennial year. Do you have any other comments or insights on engineering technology that seem to be appropriate for this year?**

Walker: I think that someone ought to do as we suggested in 1965: have a real national study, not particularly from the point of view of the engineering colleges, but go ask the people in industry what they need, and if you ask the people in industry, you find different needs. The electronics industry would have one type of thing; civil engineering would have another. But what you've got to do is find out what people have to know to get a job and hold a job. Nobody seems to attack it from that point of view.

DiGregorio: The ABET requirements, of course — you have to do that. You have to contact industry and question them and survey them. And you have to have an advisory committee, which consists of people from industry, so that's been helpful in that respect, in keeping track of what's needed, what the demand is for graduates. And I think our curriculum has changed considerably over the years based on what industry has told us. Any particular comments on ABET's effect or ABET's input into this?

Walker: ABET was always my enemy, so I've had nothing to do with them in the last 20 years.

DiGregorio: Thank you, Dr. Walker.

Appendices

Appendix A: ETD Name Changes

What's in a Name?

- 1893 SPEE
  - 1946 American Society for Engineering Education
- 1941 Technical Institute Division
  - 1971 Engineering Technology Division
- 1946 Committee of 21
  - 1962 TIC
  - 1965 TIAC
  - 1971 TCC
  - 1981 ETCC
  - 1987 ETC
- 1946 Engineering Technology Committee (ECPD)
  - 1961 ICET
  - 1980 TAC/ABET
  - 1981 NICET
- 1932 ECPD
- 1932 *Journal of Engineering Education*
  - 1969 *Engineering Education*
  - 1989 *Prism*
  - 1992 *Journal of Engineering Education*
- 1909 *Bulletin of the SPEE*

Appendix B: *TEN* Reprint

"Technical Institute Division of S.P.E.E. Has First Meeting"

A Report by President John T. Faig Ohio Mechanics Institute[1]

The newly-established Division of Technical Institutes of the Society for the Promotion of Engineering Education, which held its first conference on June 23 at the Ann Arbor meeting of that society, will provide a much needed forum for the discussion of problems encountered in this field. Most of the older technical institutes are private institutions, modestly endowed, that have been engaged in meeting definite educational needs in their respective localities. Since these needs vary from place to place, there is some tendency for technical institute courses to vary from place to place. Technical institutes have therefore developed in some measure the analysis of educational needs and the devising of courses to meet such needs that does not exist in the same degree in colleges or high schools, except in the evening classes of some colleges. There is particular need, therefore, for this forum, and it is believed that the sharing of experiences thus made possible will be of great benefit to the institutes and that the S.P.E.E. membership generally may profit by learning more about these institutions.

President W. E. Wickenden's exhaustive report on technical institutes, published about ten years ago, gives the most authoritative information about technical institutes in this country. This was later condensed by Dr. Charles F. Scott, whose experience with the Westinghouse company and later as professor of electrical engineering at Yale, fitted him splendidly for this project. Both of these reports are available.

It is an interesting fact that

---

[1] Reprinted by permission from *Technical Education News* 1, no.1 (1941): 3.

the strongest boosters of technical institutes are men in the college field who realize the need for more varied types of education than engineering colleges are fitted to offer.

The spirited paper by President P. R. Kolbe of Drexel Institute of Technology on "Presenting the Case for the Technical Institutes," which opened the conference, was read by Director W. T. Spivey, of Drexel Institute. It mentioned the difficulties encountered by technical institutes and the lack of adequate recognition of their graduates by Civil Service boards and by some personnel officers in industry. It proposed several programs for action and was altogether an excellent summing up of the situation. This paper was ably discussed by Dr. Robert L. Sackett, who has demonstrated his interest in education for industry in many ways.

President Wickenden's opening of the symposium on the place of the technical institute graduate in industry was a delightful presentation of the evidence as developed in his study of that subject. The need was approached from several angles, and the result in each instance showed that about three technical institute graduates are required for each engineering college graduate. Dr. Wickenden mentioned the development of technical institute courses in the West, particularly California, cited also interesting courses developed by industrial establishments, inspired by the defense program and the need for trained men in aviation.

The symposium continued with remarks by Dr. Dugald C. Jackson, Mr. Robert Spahr, Mr. C. C. Carr, of Pratt Institute, President A. C. Harper, of Wyomissing Institute, Dr. Charles F. Scott, Dean C. J. Freund, of Detroit University, and a number of other representatives of educational institutions and of industry.

It is understood that the committee the coming year is planning a similar [sic] for conference at the next meeting of the S.P.E.E.

Appendix C: ETD/ETC Officers

Engineering Technology Division Chair
1947-1995

1947-49	Walter L. Hughes	Franklin Institute of Boston
1949-51	Henry P. Adams	Oklahoma A & M
1951-52	Leo F. Smith	Rochester Institute of Technology
1952-53	C. L. Foster	Central Radio & TV Schools
1953-55	C. S. Jones	Academy of Aeronautics
1955-57	Karl O. Werwath	Milwaukee School of Engineering
1957-59	Kenneth L. Holderman	Pennsylvania State University
1959-61	A. Ray Sims	University of Houston
1961-63	Howard H. Kerr	Ryerson Institute of Technology
1963-65	H. C. Roundtree	Purdue University
1965-67	Richard J. Ungrodt	Milwaukee School of Engineering
1967-69	Robert J. Wear	College of Aeronautics
1969-71	Michael C. Mazzola	Franklin Institute of Boston
1971-73	Ernest R. Weidhaas	Pennsylvania State University
1973-75	Walter E. Thomas	Purdue University
1975-77	Rolf E. Davey	Wentworth Institute of Technology
1977-79	Kenneth C. Briegel	Honeywell, Inc.
1979-81	Frank A. Gourley, Jr.	North Carolina Department of Community Colleges/ Carolina Power and Light
1981-83	Gerald A. Rath	Wichita State University
1983-85	Anthony L. Tilmans	California State Polytechnic University, Pomona/Wentworth Institute of Technology
1985-87	Harris T. Travis	Southern Technical Institute
1987-89	Earl E. Gottsman	Capitol College
1989-91	George A. Timblin	Central Piedmont Community College
1991-93	Fred W. Emshousen	Purdue University
1993-95	Gary R. Crossman	Old Dominion University

Engineering Technology Council Chair
1960-1994

1960-62	Karl O. Werwath	Milwaukee School of Engineering
1962-64[a]		
1964-66	Walter L. Hartung	Academy of Aeronautics
1966-68	Lawrence V. Johnson	Southern Technical Institute
1968-70	Eugene W. Smith	Cogswell Polytechnic Institute
1970-72	Winston D. Purvine	Oregon Technical Institute
1972-74	Richard J. Ungrodt	Milwaukee School of Engineering
1974-76	Robert J. Wear	Academy of Aeronautics
1976-78	Michael C. Mazzola	Franklin Institute of Boston
1978-80	Walter O. Carlson	Southern Technical Institute
1980-82	Ernest R. Weidhaas	Pennsylvania State University
1982-84	Gary T. Fraser	SUNY-Alfred
1984-86	Edward T. Kirkpatrick	Wentworth Institute of Technology
1986-88	Ray L. Sisson	University of Southern Colorado
1988-90	Anthony L. Tilmans	Kansas Technical Institute
1990-92	Stephen R. Cheshier	Southern College of Technology
1992-94	W. David Baker	Rochester Institute of Technology

[a] Data unavailable

ETD Program Chair
1950-1969

1950-51	K. L. Burroughs	Aeronautical University
1951-53	Edward E. Booher	McGraw-Hill Book Company
1953-55[a]		
1955-57	H. Walter Shaw	McGraw-Hill Book Company
1957-58	Bonham Campbell	University of California
1958-59	R. E. McCord	Pennsylvania State University
1959-60	Lloyd W. Scholl	Purdue University
1960-61	Dana B. Hamil	Pennsylvania State University
1961-64	H. C. Roundtree	Purdue University
1964-67	Robert J. Wear	Academy of Aeronautics
1967-68	Larry J. Sitterlee	Broome Technical Community College
1968-69[b]	John R. Martin	University of Houston

[a] Data unavailable
[b] Last year as separate program chair; position changed to vice-chair, program

ETD Vice-Chair, Program
1949-1994

1949-51[a]	Karl O. Werwath	Milwaukee School of Engineering
1951-53	Louis H. Rouillion	Mechanics Institute
1953-55	K. L. Burroughs	Aeronautical University
1955-57	Kenneth L. Holderman	Pennsylvania State University
1957-59	A. Ray Sims	University of Houston
1959-61	H. W. Hartley	Northrop Institute of Technology
1961-63	H. C. Roundtree	Purdue University
1963-64	John R. Martin	University of Houston
1964-65	Richard J. Ungrodt	Milwaukee School of Engineering
1965-66	Hoyt L. McClure	Southern Technical Institute
1966-67	Lawrence J. Sitterlee	Broome Technical Community College
1967-69	John R. Martin	University of Houston
1969-71[b]	R. L. Grigsby	Richland TEC
1971-73[c]	Walter E. Thomas	Purdue University
1973-74	Rolf E. Davey	Wentworth Institute of Technology
1974-75	Ray G. Prevost	Oregon Institute of Technology
1975-76	Richard D'Onofrio	Franklin Institute of Boston
1976-77	Kenneth C. Briegel	Honeywell, Inc.
1977-78	Issac A. Morgulis	Ryerson Polytechnic Institute
1978-79	Frank A. Gourley, Jr.	North Carolina Department of Community Colleges
1979-80	James P. Todd	California State Polytechnic University, Pomona
1980-81	Donald J. Buchwald	Kansas Technical Institute
1981-82	Anthony L. Tilmans	California State Polytechnic University, Pomona
1982-83	Harris T. Travis	Southern Technical Institute
1983-84	Lawrence J. Wolf	University of Houston
1984-85	Ronald C. Paré	University of Houston
1985-86	George A. Timblin	Central Piedmont Community College

1986-87	William F. Schallert	Parks College of St. Louis University
1987-88	William S. Byers	University of Central Florida, Brevard
1988-89	Lyle B. McCurdy	California State Polytechnic University, Pomona
1989-90	Paul E. Rainey	California State Polytechnic University, San Luis Obispo
1990-91	Fred W. Emshousen	Purdue University
1991-92	Gary R. Crossman	Old Dominion University
1992-93	Albert L. McHenry	Arizona State University
1993-94	Robert English	New Jersey Institute of Technology

[a] Vice-chair only, 1949-50
[b] Vice-chair and vice-chair, program, merged in 1972
[c] Changed to one-year term

ETD Vice-Chair, Newsletter (or Equivalent) 1955-1993

1955-57[a]	Karl O. Werwath	Milwaukee School of Engineering	4/4
1957-59	K. L. Holderman	Pennsylvania State University	3/3
1959-61	A. Ray Sims	University of Houston	2/1
1961-63	Howard H. Kerr	Ryerson Institute of Technology	1/?
1963-64	H. C. Roundtree	Purdue University	?
1965-67[b]	Lawrence J. Sitterlee	Broome Technical Community College	3/3
1967-68	John R. Martin	University of Houston	3
1968-69	Michael C. Mazzola	Franklin Institute of Boston	2
1969-71[c]	Ernest R. Weidhaas	Pennsylvania State University	3/2
1971-73	Rolf E. Davey	Wentworth Institute of Technology	2/2
1973-74[d]	Ray G. Prevost	Oregon Technical Institute	1
1974-75	Richard P. D'Onofrio	Franklin Institute	2
1975-76	Kenneth Briegel	Honeywell, Inc.	2
1976-77	Issac A. Morgulis	Ryerson Polytechnical Institute	2
1977-78	Frank A. Gourley, Jr.	North Carolina Department of Community Colleges	2
1978-79	James P. Todd	California State Polytechnic University, Pomona	2
1979-80	Donald J. Buchwald	Kansas Technical Institute	2

1980-81	Anthony L. Tilmans	Indiana State University, Evansville	2
1981-82	Harris T. Travis	Purdue University	2
1982-83	Lawrence J. Wolf	University of Houston	2
1983-84	Ronald C. Paré	University of Houston	2
1984-85[e]	George A. Timblin	Central Piedmont Community College	2
1985-86	William F. Schallert	Parks College of St. Louis University	2
1986-87	William S. Byers	University of Central Florida, Brevard	2
1987-88	Lyle B. McCurdy	California State Polytechnic University, Pomona	2
1988-89	Paul E. Rainey	California State Polytechnic University, San Luis Obispo	2
1989-90	Fred W. Emshousen	Purdue University	2
1990-91	Gary R. Crossman	Old Dominion University	2
1991-92	Albert L. McHenry	Arizona State University	2
1992-93	Robert English	New Jersey Institute of Technology	2
1993-94	Warren R. Hill	Weber State College	2

[a] First newsletter published by TID chair, 1955-56. Numbers on the far right refer to number of issues per year
[b] Newsletter editor established, 1966-67
[c] Vice-chair, newsletter established, 1969-70
[d] Changed to one-year term
[e] Changed from 8 1/2" x 11" to 5 1/2" x 8 1/2" format

ETD Secretary-Treasurer
1949-1957

1949-51	Harold. T. Rodes	Ohio Mechanics Institute
1951-55	Karl O. Werwath	Milwaukee School of Engineering
1955-57[a]	Donald C. Metz	University of Dayton

[a] 1957-58 was the last year secretary and treasurer positions were together

ETD Secretary
1957-1994

1957-59[a]	Lawrence V. Johnson	Southern Technical Institute
1959-63	Eugene W. Smith	Cogswell Polytechnical Institute
1963-65	Hoyt L. McClure	Southern Technical Institute
1965-67	Paul T. Meier	Oregon Technical Institute
1967-69	R. L. Grigsby	Richland TEC
1969-71	N. B. VeerKamp	Milwaukee School of Engineering
1971-73	Ray G. Prevost	Oregon Technical Institute
1973-74	William S. Newman	Southern Technical Institute
1974-75	Rolf E. Davey	Wentworth Institute of Technology
1975-78	James P. Todd	California State Polytechnic University, Pomona
1978-80	Gerald A. Rath	Wichita State University
1980-82	Ronald C. Paré	Cogswell-North
1982-84	Durward R. Huffman	Nashville State Technical Institute
1984-86	James D. McBrayer	Franklin University
1986-88	Kenneth K. Gowdy	Kansas State University
1988-90	Linda C. Miller	St. Louis Community College at Florissant Valley
1990-92	John J. McDonough	University of Maine, Orono
1992-94	William A. Welsh	Pennsylvania State University, Harrisburg

[a] 1957-58 was the first year secretary and treasurer positions were separate

ETD Treasurer
1957-1995

1957-63[a]	Donald C. Metz	University of Dayton
1963-65	Robert J. Wear	Academy of Aeronautics
1965-67	Maurice W. Roney	Oklahoma State University
1967-69	Jerry Dobrovolny	University of Illinois
1969-71	William S. Newman	Southern Technical Institute
1971-73	William S. Newton III	Southern Technical Institute
1973-75	James V. Malone	University of Houston
1975-77	Gerald A. Rath	Wichita State University

1977-81	Amogene F. DeVaney	Amarillo College
1981-83	Rolf E. Davey	Wentworth Institute of Technology
1983-85	Thomas A. Kannema	Arizona State University
1985-86	Chris R. Conti	Merrill Publishing
	Rolf E. Davey	Wentworth Institute of Technology
1986-87	Rolf E. Davey	Wentworth Institute of Technology
1987-89	William A. Welsh	Pennsylvania State University, Capitol
1989-91	William D. Rezak	Southern College of Technology
1991-93	Alexander W. Avtgis	Wentworth Institute of Technology
1993-95	Michael T. O'Hair	Purdue University, Kokomo

[a] 1957-58 was the first year secretary and treasurer positions were separate

ETD/ETC Historian
1965-1993

1965-77	Edward L. Fleckenstein	Temple University
1977-82	Jack Spille	University of Cincinnati
1982-85	Michael T. O'Hair	Purdue University, Kokomo
1985-	Ann Montgomery Smith	Wentworth Institute of Technology

Engineering Technology Council
Summary of Officers and Directors
1970-1994

	1970-72	1972-74	1974-75	1975-76
Chair	Purvine	Ungrodt	Wear	Wear
Chair-elect				Mazzola
Secretary				Carlson
Directors				Aidala
				Curtis
				Brach
				Morgulis
				Cooke
Affiliate directors				
Representative to society nominating committee				Purvine

	_1976-77_	_1977-78_	_1978-79_
Chair	Mazzola	Mazzola	Mazzola
Chair-elect	Carlson	Carlson	Weidhaas
Secretary	Brooks	Brooks	Eklund
Directors	Ellis	Ellis	Engelson
	Goliber	Goliber	Spille
	Eklund	Eklund	Davey
	Morgulis	Dunning	Dunning
	Cooke	Michael	Michael
Affiliate directors	Seltzer	Seltzer	Katz
	Behm	Behm	Tilmans
Representative to society nominating committee	Wear	Wear	Mazzola

	_1979-80_	_1980-81_	_1981-82_
Chair	Mazzola	Weidhaas	Weidhaas
Chair-elect	Weidhaas	Fraser	Fraser
Secretary	Eklund	Sisson	Sisson
Directors	Engelson	Houston	Houston
	Spille	Todd	Todd
	Davey	Davey	Eklund
	Forman	Forman	Johnson
	Katz	Katz	Wisz
Affiliate directors	Tilmans	Seltzer	Seltzer
		Galbiati[a]	Galbiati
Representative to society nominating committee	Mazzola	Carlson	Carlson

[a] Appointed for two years but should have been one year

	_1982-83_	_1983-84_	_1984-85_
Chair	Fraser	Fraser	Kirkpatrick
Chair-elect	Todd	Todd	Sisson
Secretary	Wolf	Wolf	Gottsman
Directors	McNutt	McNutt	Caldwell
	Gorrill	Gorrill	Wolf
	Eklund	Gowdy	Cheshier

	_1982-83_	_1983-84_	_1984-85_
	Johnson	Rath	Rath
	Wisz	Tilmans	Tilmans
Affiliate directors	Haefer	Haefer	Stocker
	Timblin	O'Hair	O'Hair
Representative to society nominating committee	Weidhaas	Weidhaas	Fraser

	_1985-86_	_1986-87_	_1987-88_
Chair	Kirkpatrick	Sisson	Sisson
Chair-elect	Sisson	Tilmans	Tilmans
Secretary	Gottsman	McDonough	McDonough
Directors	Caldwell	Stevens	Stevens
	Wolf	Gottsman	Gottsman
	Cheshier	Cheshier	Faulkner
	Kanneman	Kanneman	Kryman
	Timblin	Timbin	O'Hair
Affiliate directors	Stocker	Rudnick	Rudnick
	King	King	Stocker
Representative to society nominating committee	Fraser	Kirkpatrick	Kirkpatrick

	_1988-89_	_1989-90_	_1990-91_
Chair	Tilmans	Tilmans	Cheshier
Chair-elect	Cheshier	Cheshier	Baker
Secretary	Reid	Reid	Downey
Directors	Moore	Moore	Caldwell
	Baker	Baker	Rainey
	Faulkner	Mielock	Mielock
	Kryman	Pounds	Pounds
	O'Hair	Davis	Davis
Affiliate directors	Downey	Downey	King
	Stocker	Thomas	Thomas
Representative to society nominating committee	Sisson	Sisson	Tilmans

	_1991-92_	_1992-93_	_1993-94_
Chair	Cheshier	Baker	Baker
Chair-elect	Baker	McDonough	McDonough
Secretary	Downey	Pounds	Pounds
Directors	Caldwell	Faulkner	Faulkner
	Rainey	Gourley	Gourley
	Mielock	Andrews	
	DiGregorio	DiGregorio	
	O'Hair	O'Hair	
Affiliate directors	King	Mumford	Mumford
	Rathod	Rathod	
Representative to society nominating committee	Tilmans	Cheshier	Cheshier

316   ENGINEERING TECHNOLOGY: AN ASEE HISTORY

Appendix D: ETD Newsletter Data

TID Newsletter Topics[1]
September 1955 - May 1967

Volume 1, Number 1, September 1955, 3pp.
- Toward Better Membership Communications
- Broadened Base of Activity
- Next It's Toronto
- Six Selected Subjects
- *TEN* Supplement Covers PA Meeting

Volume 1, Number 2, October 1955, 3pp.
- Toronto Hosts Announce Board Schedule
- Summary of Division Action, June 20-23, 1955

Volume 2, Number 1, February 1956, 3pp.
- New Committee Men Join in Work
- Division Membership—For Improved Organization
- Carnegie Corporation Approves Fund
- General Council Acts Favorably on Resolution
- On to Ames, Iowa—June 25-27
- Toward Improving Industrial Understanding

Volume 2, Number 2, June 1956, 5pp.
- Membership Committee Shows Action
- National Technical Institute Survey Is Launched
- 1956 Annual Meeting Preparations Complete
- TI Notes of Current Interest
- Training the Capable Minds

Volume 2, Number 3, August 1956, 4pp.
- A Report of the ASEE Committee on Degree Designations

---

[1]All newsletters between September 1955 and September 1960 included the names of the TID Executive Committee and the Committee of 21.

- Welcome to Four New Committee Members
- New Committee Structure
- The Board Meets in Detroit October 24
- America in the Engineering Age
- A Tribute to a Friend
- NCDSE Program Aids Technical Institute Education

Volume 2, Number 4, October 1956, 3pp.
- President Everitt Appoints McGraw Committee
- "The Engineering Technician" to Be Revised
[Other items covered had no headings]

Volume 3, Number 1, March 1957, 4pp.
[No headings. Included annual program of the division and a resolution in reaction to proposed federal legislation]

Volume 3, Number 2, April 1957, 4pp.
[No headings]

Volume 4, Number 1, October 1957, 3pp.
- Committees
- Dues
- Mid-Year Meeting
- A New Technical Institute Award
- *Dictionary of Occupational Titles*
- A Retirement
- Annual Meeting

Volume 4, Number 2, March 1958, 4pp.
- Summary of the Mid-Year Meeting of the National Committee
- From the Chairman
- Annual Meeting
- The Survey
- Summer Institutes
- Recent Announcements of Interest
- News Notes

Volume 4, Number 3, May 1958, 4pp.
- From the Chairman

- The Annual Meeting
- Relations with Government
- "The Engineering Technician"
- Summer Institute on Nuclear Technology
- News Notes of Interest
- Committee Interest Survey Form

Volume 5, Number 1, September 1958, 2pp.
 [No headings]

Volume 5, Number 2, January 1958, 4pp.
- Evaluation of Technical Institute Education
- National Defense Education Act
- Teacher Exchange Program
- Summer Institutes
- Miscellaneous News Notes

Volume 5, Number 3, April 1959, 3pp.
- Summer Institutes
- National Interest
- The "Henninger Project"
- Foreign Exchange
- An Evaluation of TI Education
- Random Notes

Volume 6, Number 1, September 1959, 4pp.
- Greetings from the Chairman
- NSF Institute
- NSF-FIER Conference
- NSF to Sponsor Two Institutes in 1960
- Daytona Beach Junior College to Begin Technical Institute
- Standing Committees

Volume 6, Number 2, February 1960, 4pp.
- NSF Schedules Two Institutes for Technical Institute Faculties
- Teacher Exchanges with Great Britain Announced
- Program Completed for Purdue Meeting

Volume 7, Number 1, September 1960, 6pp.
- Mid-Year Meeting

- June Meeting
- English-Speaking Union Fall Faculty Exchange
- Summer Institutes for TI Faculty Completed
- Mailing List Being Rectified
- Definitions Approved
- Standing Committees

January 1962, 5pp.
- Annual Meeting
- A Reorganization Plan Submitted by the Long-Range Planning Committee
- Teacher-Exchange Program
- Technical Institute Evaluation
- A. L. Williston Award
- ASEE/ASTME One Day Summer School
- Editor
- Standing and *ad hoc* Committee Chairmen
- Nominating Committee
- NSF Summer Institutes for Engineering Teachers
- "The Engineering Technician"
- James H. McGraw Award

First Issue, December 1966, 6pp.
- Chairman's Corner
- From the Editor
- ECPD Accreditation for 4-Year Programs
- NSF Rejects Goals Study; Second Try Planned
- Special ASEE Journal
- TID Committee Activities
- Housing for Members
- New Booklet
- McGraw Award

Volume 1, Number 2, May 1967, 8pp.
- Chairman's Corner
- Faculty Organize Southeastern Section of TID
- Did You Know That...
- New Curricula in September
- From the Editor
- Program for Annual Meeting

Topics Appearing Only Once in the ETD Newsletter
February 1968-Spring 1992

February 1968	· The Baccalaureate Degree
September 1969	· Jobs and Black Students
January 1970	· Disadvantaged
May 1970	· Agricultural Equipment Technology
May 1974	· Science and Engineering Technology Education Program
December 1974	· Council News
Fall 1976	· Metrication Coordination Committee
	· NSPE
Fall 1980	· New Name for ECPD
	· ABET Report
Spring 1981	· DELOS Award
	· Cross-Tabulation of Membership
	· ETD *ad hoc* Committees
	· ASEE Chapters
	· Committee on Faculty Concerns
Fall 1981	· Program of Work Committee
Spring 1982	· Resolutions
	· IEEE Position on ET
Fall 1984	· Fellow Grade Nominations
Spring 1985	· NSF Report on Engineering Technology
Fall 1985	· ET College Listing
Spring 1987	· Viewpoints from an International Student
	· Ride the Rails to Reno
	· ETD Membership List
Fall 1987	· National Forum for Engineering Technology
Spring 1988	· ET Articulation Agreement
	· Graduate Programs
Fall 1988	· ET Research
	· ETC Membership
Spring 1989	· Survey of Textbooks and Lab Manuals
	· Cooperative Education Division Program
Fall 1989	· ETD Brochures
	· Computer Software
	· Editor/Reviewers

Spring 1991 · ET Calendar
· ET Centennial Committee
· Directory of ET Institutions and Programs
· ETD Mission Statement
Spring 1992 · ETC International Committee

ETD Newsletter Topics
1967-1992

	12/67 7pp	2/68 9pp	5/68 10pp	11/68 4pp	3/69 5pp	9/69 13pp
Annual conference report	■				■	■
Technifacts	■	■	■	■	■	■
News from institutions	■					
Section activities/representatives			■			
Annual conference program	■		■			
Chairman's corner				■	■	
Editorial/editor's notes						■
Study of ET education						■
Division officers/committee chairs						■

	1/70 11pp	5/70 9pp	9/70 13pp	2/71 6pp	10/71 15pp	5/72 7pp
Annual conference report	■	■	■		■	
Technifacts	■	■	■	■	■	■
News from institutions	■					
Annual conference program					■	■
Chairman's corner				■	■	■
Editorial/editor's notes	■	■	■	■		
Study of ET education	■	■				■
ICET/ASCET/NICET			■			
Division officers/committee chairs	■	■	■	■	■	■
ETC business meeting minutes	■					
ET brochures	■				■	
International cooperation	■					
ETD membership			■		■	■

	1/70 11pp	5/70 9pp	9/70 13pp	2/71 6pp	10/71 15pp	5/72 7pp
ETD business meeting minutes			■		■	
McGraw Award				■		
Division by-laws				■		
Leadership makes a difference				■	■	

	12/72 13pp	5/73 9pp	5/74 19pp	12/74 15pp	5/75 13pp	F/75 9pp
Annual conference report						■
Technifacts	■	■				
News from institutions			■	■	■	
Annual conference program	■		■	■		
Chairman's corner	■	■	■	■		
Editorial/editor's notes	■					■
ICET/ASCET/NICET	■					■
Division officers/committee chairs	■	■		■	■	
ETD membership	■	■		■		
ETD business meeting minutes	■		■	■		
McGraw Award		■				
ETD by-laws	■					
ETD annual report	■					
ETC annual report/news			■	■		
Historian/ET archives			■			
Publications			■			
ETLI			■			
ETC Manpower Committee/EMC			■			■
Tau Alpha Pi news				■		
Magnum opus; articles, letters				■	■	
CIEC						■

	S/76 8pp	F/76 7pp	S/77 10pp	F/77 15pp	S/78 6pp	F/78 9pp
Annual conference report			■			
News from institutions			■	■	■	■
Annual conference program	■		■	■		■
Chairman's corner	■	■	■	■	■	■
ICET/ASCET/NICET			■			
Division officer/committee chairs		■			■	
ETD membership	■	■	■			
ETD business meeting minutes		■			■	■
ETD by-laws				■		
ETD annual report		■				
Historian/ET archives				■		
Publications				■		
ETLI						■
TC Manpower Committee/EMC					■	
Tau Alphi Pi News	■		■			
CIEC	■	■	■	■	■	
ETC Publications Committee	■					
SME	■					
ETD Newsletter Committee					■	■
ET definitions/Development Committee/11 principles					■	
Conference workshop						■

	S/79 14pp	F/79 12pp	S/80 18pp	F/80 24pp	S/81 20pp	F/81 30pp
News from institutions	■	■	■	■	■	■
Section activities/representatives		■	■	■	■	■
Annual conference program			■	■	■	
Chairman's corner	■	■	■	■	■	■
Division officers/committee chairs		■	■	■	■	■
ETC business meeting minutes	■					
ETD membership		■	■			

	S/79 14pp	F/79 12pp	S/80 18pp	F/80 24pp	S/81 20pp	F/81 30pp
ETD business meeting minutes		■	■	■	■	■
ETD by-laws	■					
ETLI	■	■			■	■
Tau Alpha Pi news	■					
CIEC		■		■	■	■
ETD Newsletter Committee	■					
ET definitions/Development Committee/11 principles		■				
Conference workshop	■	■			■	■
Call for papers, *Eng. Education*		■				
ETD mini-grants				■	■	
Dept. Heads Association					■	■
News of members					■	
ABET criteria					■	■
ET professional society						■
Position announcement						■

	S/82 15pp	F/82 15pp	S/83 14pp	F/83 13pp	S/84 11pp	F/84 31pp
News from institutions	■	■	■	■	■	■
Section activities/representatives	■	■				■
Annual conference program	■	■				■
Chairman's corner	■	■	■	■		■
Editorial/editor's notes		■	■	■	■	■
ETD officers/committee chairs		■		■	■	■
ETC business meeting minutes						■
ET brochures	■					■
ETD membership		■				■
ETD business meeting minutes		■	■	■		■
McGraw Award						■
ETC annual report	■					
ETLI		■	■	■	■	■

	S/82 15pp	F/82 15pp	S/83 14pp	F/83 13pp	S/84 11pp	F/84 31pp	S/85 27pp	F/85 42pp	S/86 22pp	F/86 32pp	S/87 54pp	F/87 31pp
Tau Alpha Pi news			■		■							
Magnum opus; articles, letters	■	■										
CIEC		■	■	■	■	■						
ETC Publications Committee						■						
SME			■		■							
ET definitions/Development Committee/11 principles		■			■							
Conference workshop	■			■	■							
Call for papers, *Eng. Education*						■						
ETD mini-grants	■	■			■	■						
News of members				■		■						
ET professional society	■											
ASEE society committees						■						
JET	■	■	■	■	■	■						
News from institutions							■	■	■		■	
Section activities/representatives							■		■		■	
Annual conference program								■		■	■	
Chairman's corner							■	■	■	■	■	■
Editorial/editor's notes							■	■	■	■	■	■
ETD officers/committee chairs							■	■	■	■	■	■
ETC business meeting minutes										■	■	
ETD membership							■		■			■
ETD business meeting minutes							■	■	■	■	■	
Leadership makes a difference							■					
ETD annual report											■	
ETC annual report								■				
Historian/ET archives										■		
ETLI							■		■			■
ETC Manpower Committee/EMC												■

	S/85 27pp	F/85 42pp	S/86 22pp	F/86 32pp	S/87 54pp	F/87 31pp
Tau Alpha Pi news	■	■	■		■	
CIEC	■	■		■		■
Call for papers, *Eng. Education*		■	■	■		■
ETD mini-grants	■	■	■			
News of members	■					■
ABET criteria	■					
JET	■	■	■			
QEEP	■		■			
SIG's	■	■				
Call for papers, annual conference	■	■				
Sparks Award (ASME)				■		
World Congress on ET Education				■		
New ETD members						■

	S/88 23pp	F/88 31pp	S/89 29pp	F/89 19pp	S/90 23pp	F/90 17pp
News from institutions		■	■	■	■	■
Section activities/representatives			■	■	■	■
Annual conference program	■			■	■	■
Chairman's corner	■	■	■	■	■	■
Editorial/editor's notes	■	■	■			
ICET/ASCET/NICET	■					
ETD officer/committee chairs	■	■	■	■		■
ETC business meeting minutes	■	■			■	
ET brochures				■		
ETD membership	■					
ETD business meeting minutes	■					
McGraw Award	■	■				
Historian/ET archives			■			
Publications	■		■	■		
ETLI	■		■		■	
CIEC		■		■		

	S/88 23pp	F/88 31pp	S/89 29pp	F/89 19pp	S/90 23pp	F/90 17pp
ETC Publications Committee			■			
SME	■					
Call for papers, *Eng. Education*	■	■				
News of members	■	■	■			■
ABET criteria				■	■	■
JET				■	■	■
Call for papers, annual conference	■			■	■	■
Sparks Award (ASME)					■	■
World Congress on ET Education	■					
New ETD members	■	■	■	■		
ET bibliography			■	■		
10-volume compendium		■		■		■
Peer review process		■	■	■		
Berger Award	■					

	S/91 19pp	F/91[a]	S/92 29pp
News from institutions	■		■
Chairman's corner	■		
ETD business meeting minutes	■		■
ETLI	■		■
ET definitions/Development Committee/11 principles	■		
ETD mini-grants	■		
News of members	■		
JET	■		
Call for papers, annual conference	■		■
New ETD members	■		
10-volume compendium	■		
Berger Award	■		

[a] Data unavailable

Appendix E: ETLI

ETLI Host Institutions and Executive Council Chairs
1976-1992

Year	Institution	Chair
1976	Indiana State University, Evansville	Anthony L. Tilmans
1977	Purdue University	Samuel L. Pritchett
1978	Purdue University	Stephen R. Cheshier
1979	Nashville State Technical Institute	Boyce D. Tate
1980	Southern Technical Institute	Anthony L. Tilmans
1981	Arizona State University	Stephen R. Cheshier
1982	University of Houston	Stephen R. Cheshier
1983	Wentworth Institute of Technology	Thomas A. Kanneman
1984	California State Polytechnic University, Pomona	Thomas A. Kanneman
1985	Southern Technical Institute	Ray L. Sisson
1986	Capitol Institute of Technology	W. David Baker
1987	Nashville State Technical Institute	Anthony L. Tilmans
1988	Purdue University	John J. McDonough
1989	Old Dominion University	Rolf E. Davey
1990	Kansas College of Technology	Steven W. Faulkner
1991	Pennsylvania State University, Harrisburg	William D. Rezak
1992	Oregon Institute of Technology	Warren R. Hill

Appendix F: *JET* Data

JET Editorial Board Members
1983-1995

Editor-in-Chief

1983-85	Kenneth G. Merkel	University of Nebraska, Omaha
1985-87	Lawrence J. Wolf	University of Houston
1987-89	Michael T. O'Hair	Purdue University, Kokomo
1989-91	Durward R. Huffman	Northern Maine Technical College
1991-93	Elliot Eisenberg	Pennsylvania State University, Hazleton
1993-95	Carole E. Goodson	University of Houston

Production Editor

1983-85	Lawrence J. Wolf	University of Houston
1985-87	Michael T. O'Hair	Purdue University, Kokomo
1987-89	Durward R. Huffman	Northern Maine Technical College
1989-91	Elliot Eisenberg	Pennsylvania State University, Hazleton
1991-93	Carole E. Goodson	University of Houston
1993-95	W. Frank Reeve	Purdue University

Editorial Assistant

1983-87	Barbara A. Wolf	Humble, Texas

Advertising Editor

1987-89	C. Howard Heiden	University of Southern Mississippi
1989-91	James D. McBrayer	Franklin University
1991-93	W. Frank Reeve	Purdue University
1993-95	Richard M. Moore	Oregon Institute of Technology

Manuscript Editor

1987-89	Elliot Eisenberg	Pennsylvania State University, Hazleton
1989-91	C. Howard Heiden	University of Southern Mississippi
1991-92	James D. McBrayer	University of Central Florida
1992-93	Richard M. Moore	Oregon Institute of Technology
1993-95	Cecil Harrison	University of Southern Mississippi

Subscription Editor

1987	Ronald S. Scott	Northeastern University
1987-88	Durward R. Huffman	Nashville State Technical Institute
1989-	Thomas J. Bingham, Jr.	St. Louis Community College at Florissant Valley

At-Large Members

1983-85	Michael T. O'Hair	Purdue University, Kokomo
1983-87	Durward R. Huffman	Nashville State Technical College
1983-87	Ronald S. Scott	Wentworth Institute of Technology
1985-87	Thomas A. Kanneman	Arizona State University

Past Editor

1985-87	Kenneth G. Merkel	University of Nebraska, Omaha
1987-89	Lawrence J. Wolf	University of Houston
1989-91	Michael T. O'Hair	Purdue University, Kokomo
1991-93	Durward R. Huffman	Northern Maine Technical College
1993-95	Elliot Eisenberg	Pennsylvania State University, Hazleton

Appendix G: ETD Meeting Minutes

Committees and Reports
1949-1991

Year beginning:	1949	1950	1951	1952	1953	1954
TI teacher training	■	■	■	■		■
Completion credentials	■	■	■	■		■
Cooperation with government agencies	■[a]		■[a]		■[a]	■
Curriculum development	■		■		■	■
Membership	■	■				
Student selection, guidance	■	■				
Program	■	■	■			■
Committee of 21	■	■	■	■	■	■
Nominations	■	■	■	■	■	■
TI studies (editor)		■	■		■	■
Relations with industry			■		■	■
Manpower studies			■		■	■
Relations with professional societies			■		■	■
Degree designation				A		
McGraw Award					■	■
Place of general education						■

Year beginning:	1955	1956[b]	1960	1961	1962	1963
TI teacher training	■	■	■[c]	■[c]		
Completion credentials	■					
Cooperation with government agencies	■	■	■	■		
Curriculum development	■	■	■	■		
Membership		■	■	■	■	
Student selection, guidance	■	■	■	■	■	
Program		■	■		■	■
Committee of 21	■	■	■	■	■	■
Nominations	■	■	■	■		

Year beginning:	1955	1956[b]	1960	1961	1962	1963
TI studies (editor)	■[d]	■[d]	■[d]			
Relations with industry	■	■	■	■		
Manpower studies	■	■				■
Relations with professional societies	■	■	■			
Degree designation	■					
McGraw Award	■	■	■	■	■	■
Place of general education	■	■	■	■		
Study fund	■	■				
Metals technology	■	■				
Nucleonics	■	■				
Relations with educational organizations				■	■	
Relations with high schools				■	■	■
Division reorganization				■	■	
TIAC/TCC					■	■
Williston Award					■	■
ETD by-laws					■	■
ETLI					■	■

Year beginning:	1964	1965	1966	1967	1968	1969
Cooperation with government agencies	■					
Membership		■	■		■	■
Program			■	■	■	
Nominations			■	■	■	
Manpower studies	■		■			
Relations with professional societies	■					
McGraw Award				■		
Relations with educational organizations	■					

Appendices **333**

Year beginning:	*1964*	*1965*	*1966*	*1967*	*1968*	*1969*
TIAC/TCC/ETCC/ETC				■	■	
Historian	■	■	■			
Section representatives			■	■	■	■

Year beginning:	*1970*	*1971*	*1972*	*1973*	*1974*	*1975*
Membership	■	■	■	■	■	■
Program						■
Nominations	■	■	■	■	■	■
TIAC/TCC	■				■	■
ETD by-laws			■			
Historian			■			
SIG's			■			

Year beginning:	*1976*	*1977*	*1978*	*1979*	*1980*	*1981*
Membership	■	■	■		■	■
Program	■	■	■			■
Nominations	■	■	■	■	■	■
ETCC	■	■	■		■	■
Historian		■	■			■
Section representatives				■	■	■
Mini-grants					■	■
CIEC	■	■	■	■		■
Program of work						■

Year beginning:	*1982*	*1983*	*1984*	*1985*	*1986*	*1987*
ET brochure					■	
Membership	■	■	■	■		■
Program	■	■	■	■	■	■
Nominations	■		■	■		
McGraw Award				■	■	
ETC	■	■		■	■	■
By-laws			■			

Year beginning:	1982	1983	1984	1985	1986	1987
Historian	■	■		■	■	■
Section representatives	■	■		■	■	■
SIG's			■			
Mini-grants	■	■	■	■	■	■
CIEC	■	■	■	■	■	■
JET	■	■	■	■	■	■
Program of work	■	■				
ET brochure		■	■	■	■	
Development Committee	■	■	■			
Goals and activities			■		■	■

Year beginning:	1988[a]	1990	1991
Membership		■	■
Program	■	■	■
Nominations		■	■
Historian		■	■
Section representatives		■	■
Mini-grants		■	■
Goals and activities	■	■	
Centennial			■

[a] Data unavailable for 1957-59, 1989
[b] Cooperation with Office of National Defense
[c] Identified as editor
[d] Teacher training and recruitment
A Appointed

Appendix H: Membership Data

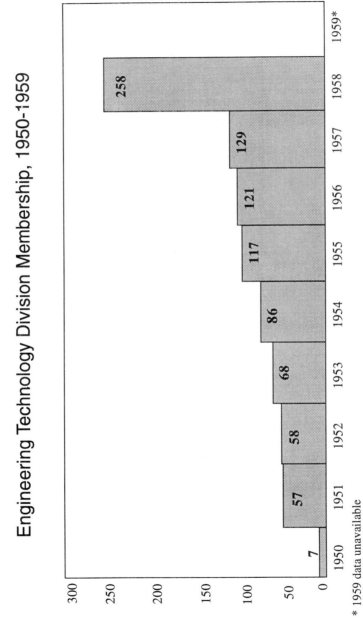

Engineering Technology Division Membership, 1950-1959

Year	Members
1950	7
1951	57
1952	58
1953	68
1954	86
1955	117
1956	121
1957	129
1958	258
1959*	

* 1959 data unavailable
** Membership numbers taken by headcount, 1950-1959

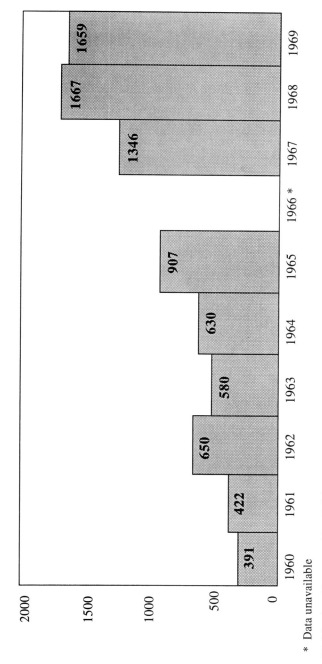

Appendices **337**

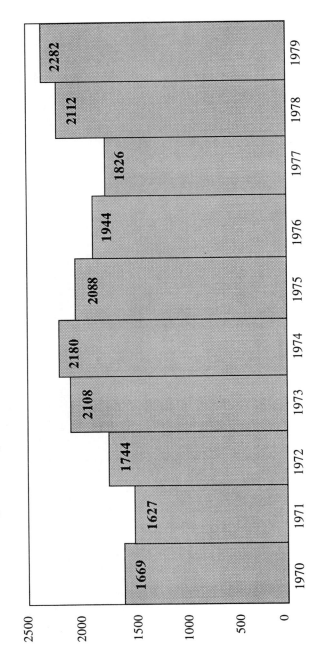

Engineering Technology Division Membership, 1970-1979

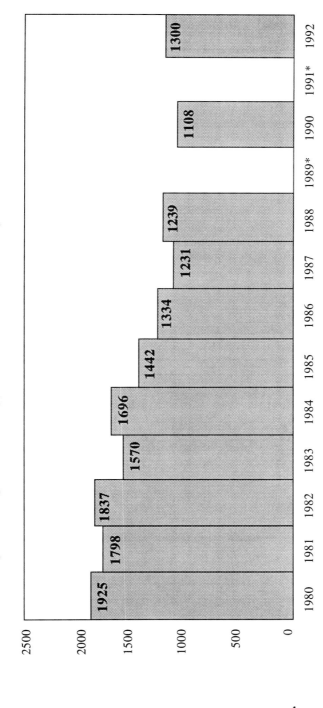

Appendices **339**

Appendix I: Captions and Credits, Photos

Front Cover

1	2	3
4	5	6
7	8	9
10	11	12

1. Professor Howard Dyche, EET Group Leader, checks progress of his McKeesport Seniors...photo courtesy of Penn State

2. The Journal of Engineering Technology in production...photo by Barbara Wolf

3. Surveying at Southern...photo courtesy of Southern College of Technology

4. Choosing a cover color for the March 1985 Journal of Engineering Technology...Photo by Barbara Wolf

5. 1955, first woman technology faculty member, Jill McCormick, joins the aviation technology faculty at Purdue University...photo courtesy of Purdue University

6. CIM cell fires up at OIT, 1992...photo courtesy of Oregon Institute of Technology

7. Machine shop, 1950's, OIT...photo courtesy of Oregon Institute of Technology

8. Tau Alpha Pi National Honor Society, Zeta Gamma Chapter, Texas A&M, June of 1992 -- Dr. Frederick J. Berger with awardees and advisors...photo courtesy of Dr. Frederick J. Berger

9. In the Shop at Southern...photo courtesy of Southern College of Technology

10. Laser lab light show at OIT, 1992...photo courtesy of Oregon Institute of Technology

11. At work on an engine at the Casey Jones School of Aeronautics...photo courtesy of Lorretta Bailey, College of Aeronautics

12. 1993, Eddie Hildreth, Southern University, is one of the recipients of the Centennial Award...photo by Barbara Wolf

Back Cover

1   2

3   4

McGraw Award Recipients:

1. Left to right, K. Holderman, Karl Werwath, F. C. Lindvall (Karl Werwath, recipient, June 1958)...photo courtesy of Robert J. Wear

2. Harris Travis (recipient 1988) at podium during Centennial Award ceremony, 1993 ...photo by Barbara Wolf

3. Lawrence J. Wolf (recipient 1987) goes over data on students' senior project at OIT, 1993 ...photo by Barbara Wolf

4. L. V. Johnson receives McGraw Award, 1963 (pictured here with Winston Purvine, Glenn Murphy, and H. Kerr)...photo courtesy of Robert J. Wear

Chapter 1, p 6

1	2
3	4
5	6

1. Sketch for Wentworth Institute of Technology Campus, Boston, Massachusetts...photo courtesy of Wentworth Institute

2. Left to right, Lee Warrendes, Casey Jones, George Vaughn, Walter Hartung, College of Aeronautics...photo courtesy of Lorretta Bailey, College of Aeronautics

3. Frank Gourley, author of this chapter (and compiler of Chapter 5)...photo courtesy of Frank Gourley

4. Wentworth Institute of Technology, Opening Day (circa 1911)....photo courtesy of Mary Ellen Flaherty, Wentworth Institute of Technology

5. Capitol College, Laurel, Maryland, in the 1930's...photo courtesy of Capitol College

6. After completing coursework in Armature Winding at Milwaukee School of Engineering, this student went on to become an electrical inspector for the City of Los Angeles...photo courtesy of Kent Peterson, MSOE

Chapter 2, p 36

1

  2

3

1. Tony Tillmans, author of this chapter...photo courtesy of Frank Gourley

2. A meeting at Leo Ruth's home...photo by Barbara Wolf

3. Fred Emshousen...photo courtesy of Frank Gourley

Chapter 3, p 44

1   2

3   4

5   6

1., 2.,
5., 6.   The Journal of Engineering Technology in production...photos at Any-Type Any-Tyme and Hearn Lithographing Company courtesy of Barbara Wolf

3. Inaugural Issue, the Journal of Engineering Technology...photo by Mary Smothers

4. Mike O'Hair and Larry Wolf, authors of this chapter...photos courtesy of Frank Gourley

Chapter 4, p 58

    1

2   3

    4

1. R. Van Houten, H. Kerr, and Eugene Rietzke (McGraw Award recipient), 1962...photo courtesy of Robert Wear

2. E. W. Smith (recipient) and R. J. Ungrodt walk to McGraw ceremonies, 1976...photo courtesy of Robert Wear

3. Robert Wear, author of this chapter...photo courtesy of Frank Gourley

4.	H. Russell Beatty receives McGraw Award, 1961...photo courtesy of Robert Wear

Chapter 4, pp 61-100

McGraw Award recipients...photos courtesy of Robert J. Wear, Frank Gourley, ETD, Carole E. Goodson, and recipients

Chapter 5, p 102, photos courtesy of Frank Gourley

1   2

  3

4   5

1.	Gary Crossman of Old Dominion University

2.	W. David Baker of Rochester Institute of Technology

3.	ETD Business Luncheon photo, 1986 CIEC, New Orleans, Louisiana... photo by Herb Moore, courtesy of Frank Gourley

4.	Ashok Agrawal of St. Louis Community College at Florissant Valley

5.	Robert Mott of University of Dayton

Chapter 6, p 192

1   2

3

4   5

1. Casey Jones "in flying costume and...today," as he appeared in 1941 in "The Wind Tunnel," the yearbook of the Senior Engineering class of the Casey Jones School of Aeronautics...photo and booklet by Arthur Studios, New York City, courtesy of Lorretta Bailey

2. Strain gauging a valve at University of Houston...photo courtesy of Carole E. Goodson

3. Alexander Avtgis, author of introduction, this chapter...photo courtesy of Frank Gourley

4. Engine lathe operation at OIT...photo courtesy of Oregon Institute of Technology

5. The Radio and Television Technician curriculum at Milwaukee School of Engineering was designed to provide students with a technical and business background to enable them access to the rapidly expanding TV service field of the 1950's...photo courtesy of Kent Peterson

Walter M. Hartung, pp 196-197, photos courtesy of Lorretta Bailey, College of Aeronautics

1		3	4
2		5	
		6	

1. Walter M. Hartung

2., 3. Classes at the Casey Jones School of Aeronautics

4. Walter Hartung on the top of the roof of (then) Academy of Aeronautics

5. Engine Shop, School of Aeronautics

6. With one of the airplanes, Academy of Aeronautics

Lawrence V. Johnson, pp 202-203, photos courtesy of Southern College of Technology

1		3	4
2			6
		5	
			7

1. Lawrence V. Johnson

2. Students in a then state-of-the art laboratory (circa early 50's), Southern College of Technology

3. Preparations for Southern Tech's Bathtub Race

4. Student body president Jeff Tucker welcomes Governor Jimmy Carter to the Southern College of Technology campus. Between them is third director Walter Carlson, with his predecessor Hoyt McClure looking on.

5. A meeting on the Chamblee campus (first home of Southern Tech), presided over by George Crawford, first dean of the Technical Division (circa late 40's or early 50's)

6. Using a surveying level at Southern Tech

7. Southern Technical Institute's "Ham" Radio Station (circa early 50's)

Michael C. Mazzola, pp 212-213, photos courtesy of Franklin Institute of Boston and Brian Kenny, Boston University

1		3	4
2		5	

1. Michael C. Mazzola

2. Mechanical Engineering Lab (circa 1925)

3. Machining Lab (circa 1940)

4. Automotive Program (circa 1912)

5. Telegraph Lab -- to train servicemen bound for Europe in World War I -- Government contract

Hugh E. McCallick, pp 220-221, photos courtesy of Carole E. Goodson, University of Houston

1		3	4
2		5	
		6	7

1. Hugh E. McCallick

2. A Surveying class at the University of Houston

3. U of H College of Technology early in the computer graphics age

4. Professor Carole E. Goodson in Math learning laboratory

5. The College of Technology building at U of H

6. Professor Curtis Johnson with participants in the Algerian Project

7. Professor K. C. Ting with solar collector student project

George W. McNelly, pp 236-237, photos courtesy of Michael T. O'Hair and Fred Emshousen, Purdue University

1		3	4
2		5	
		6	7

1.   George W. McNelly

2.   A drafting laboratory in the Engineering Building at Purdue (circa 1912)

3.   The first Technical Extension class taught under the auspices of the Technical Institute

4.   The original facility for the Purdue Programs at Kokomo when classes began in 1968 -- the "Carriage House" of the Seiberling Mansion, the building in which the technology classes were taught (photo circa 1945)

5.   Founding staff of Purdue's newly formed School of Technology, July 1, 1964. Left to right: Denver Sams, Harry Belman, Helen Johnson, George McNelly, James Maris, Charles Hutton, Gil Rainey. In the foreground: Charles Lawshe, dean

6.   Knoy Hall at Purdue University, West Lafayette, Indiana

7.   Welding laboratory at Purdue (circa 1930)

Winston D. Purvine, pp 258-259, photos courtesy of Oregon Institute of Technology

1		3	
2		4	5
		6	

1. Winston D. Purvine

2. EET students at OIT disemboweling equipment (circa 1988)

3. A Laser Optics Engineering Technology Program lab at Oregon Institute of Technology

4. The first computer at OIT (circa 1950's), which required punched card entry

5. Early gunsmithing lab out of which the OIT Manufacturing Engineering Technology Program grew

6. First OIT campus at mile-high Marine Rehabilitation Center (circa 1947)

Richard J. Ungrodt, pp 270-271, photos courtesy of Kent Peterson, Milwaukee School of Engineering

1		3	4
2		5	
		6	

1. Richard J. Ungrodt

2. A Telephone Exchange laboratory at Milwaukee School of Engineering (circa 1920's)

3.   MSOE Founder, Oscar Werwath in an electric automobile (circa 1911)

4.   MSOE students learning to do residential electrical wiring in this two-story Model Wiring House

5.   Richard J. Ungrodt (center) as a student at Milwaukee School of Engineering (circa the late 1930's)

6.   Students in the Aerotechnician program, assembling a "B-G" glider in the MSOE aircraft laboratory (circa 1942)

Eric A. Walker, pp 292-293, photos courtesy of Dr. Wayne R. Hager, Pennsyvania State University

```
1           3   4

2               6
            5
                7
```

1.   Eric A. Walker

2.   Assistant Director Sargent and District Administrator Tomm shown looking over the plans of the new Shenango Valley campus of Penn State (circa 1965)

3.   Robert Dawson, director of the Scranton campus of Penn State, pointing out the location of the first building at the new campus site to Ernie Weidhaas, Merritt Williamson, and Fran Hall

4.   Three EET seniors pondering a problem at York. Left to right: R. Null, N. Splain, T. Sprecher

5.   Mr. McGilvra, Engineering instructor at the York campus of Penn State, helping EET student

6.   Twins Richard and Robert Seibel in the lab at Penn State

7.   A drafting laboratory at the Allentown campus of Penn State

Index

Academy of Aeronautics xi, 59, 196-201, 225, 307. See also Casey Jones School of Aeronautics; College of Aeronautics

Accreditation Advisory Service Committee 158, 161, 165

Accreditation Board for Engineering and Technology (ABET) ix, 8, 31, 91, 136, 166, 200, 277, 321

AccreditationRelations Committee 183

Adams, Henry P. 69, 104, 307

Aeronautical University 225, 308

Aidala, Joe 129, 313

Air Force Academy 123

Alabama A & M University 38

Albany College 256

Amarillo College 312

American Biarritz University 72, 204

American Institute of Aeronautics and Astronautics (AIAA) 74, 91

American Society of Certified Engineering Technicians (ASCET) 154, 322

American Society of Civil Engineers (ASCE) 88, 187, 212

American Society of Mechanical Engineers (ASME) 31, 151, 328

Antrim, John D. 21, 38

Argonne National Laboratory 119

Arizona State University 45, 309, 329

Arthur Williston Award 63

Arthur Williston Award Committee 124

Association of Computing Machinery (ACM) 149

Atomic Energy Commission 119

Avtgis, Alexander W. 151, 193, 312

Awards Committee 175

Baird, R. E. 160

Baker, W. David 21, 179, 308

Bannerman, James W. 21

Beasley, Jack xi, 270-290

Beatty, H. Russell 33, 71, 112, 200

353

Beese, Charles W. 64, 239, 240

Bell and Howell School 85

Ben Sparks Award 151

Berger Award 328

Berger, Frederick J. 3, 98, 131

Berleth, Francis 226

Bezbatchenko, Mike 38

Biedenbach, Joseph M. 21

Bingham, Thomas J., Jr. 331

Blumenthal, Leon H. 160

Booher, Edward E. 59, 105, 308

Brademas, John 249

Briegel, Kenneth C. 307, 310

Brigham Young University 38, 228

Brodsky, Stanley M. 42, 147, 172

Broome Technical Community College 75, 308. See also New York State Institute of Applied Arts and Sciences

Brown, David W. 14, 29

Brunner, Ken 122

Buchwald, Donald J. 136, 147, 309

Buczynski, Robert J. 15, 29, 145

Burdick, Glenn A. 21

Burris, Frank E 33

Burroughs, K. L. 308, 309

By-Laws Committee 19, 125, 172, 333, 335

Byers, William S. 21, 142, 309

California Institute of Technology 81, 296

California State Polytechnic University, Pomona 14, 174, 309

California State Polytechnic University, San Luis Obispo 289, 309

Callis, Charles 38

Campbell, Bonham 308

Cannon, Frank V. 21

Capitol College 1, 97, 199

Capitol Institute of Technology 182, 329

Capitol Radio Engineering Institute 72, 108

Capitol Tech 199

Capitol Trades 225

Carlson, Walter O. 134, 164, 308

Carnegie Foundation 12, 25, 113

Carnegie Institute of Technology 95

Carr, C. C. 306

Carroll, Frank T. 21

Case Institute of Technology 118

Casey Jones School of Aeronautics 72, 85, 198. See also Academy of Aeronautics; College of Aeronautics

Cavanaugh, Bill 115

Centennial Committee ix, x, xi, 150, 177, 193, 321

Central Piedmont Community College 99, 307, 309

Central Radio & TV Schools 307

Central Technical Institute 74, 199

Chenea, Paul 241

Cheshier, Stephen R. 41, 90, 149, 180, 308

Chicago Technical College 225

City University of New York 97

Civil Service Commission 3, 114, 133, 306

Coe College 234

Cogswell College 83

Cogswell Polytechnic Institute 308

Cogswell Polytechnical College 5

Cogswell Polytechnical Institute 312

Coleman, A. P. 122

College Industry Education Conference (CIEC) 20, 133, 172, 281, 323, 327

College of Aeronautics xi, 67, 307. See also Academy of Aeronautics; Casey Jones School of Aeronautics

College-Industry Council 177

College-Industry Partnerships Division 20

College-Industry Relations Committee 158

Collins, W. L. 115, 122

Colorado State University 96, 133

Committee for the Eleventh Issue 154

Committee of 21 9, 80, 118, 198, 258, 332

Committee on Engineering Education Studies 140

Committee on Engineering Technology Research 181

Committee on Minorities and Females in Engineering 94

Committee on Reorganization of the Division 123, 124

Committee on Society Awards 189

Committee on the Education and Utilization of the Engineer 181

Committee on the Place of General Studies 12, 111, 332

Communications Committee 43

Completion Credentials Committee 106, 332

Computer Orientation in Engineering Technology Committee 158

Computer Software Committee 156

Conti, Chris R. 312

Continuing Engineering Studies Division 20

Continuing Professional Development Division 20

Cooper, Bill L. 21

Cooper Union 63

Cooperation with Government Agencies Committee 109, 332

Cooperation with Office of National Defense Committee 106

Cooperative Education Division 20

Cornell University 114, 115, 117

Council Historian Committee 158

Crossman, Gary R. 151, 307, 309

Cullen, H. R. 226

Curriculum Development Committee 11, 106, 121, 332

Daniel Guggenheim School of Aeronautics 72, 202

Davey, Rolf E. 134, 167, 170, 178, 307, 312

Definitions Committee 164

Defore, Jesse J. 26, 132, 208

Delgado College 38

Detroit University 306

DeVaney, Amogene F. 312

Development Committee 335

DeVry Institute 5, 199, 200

Dictionary of Occupational Titles 109, 318

DiGregorio, Joseph xi, 186, 315

Dilly, Ron 53

Directory of Engineering Technology Institutions 18, 150, 186

Dobbs, Fredrick E. 66

Dobrovolny, Jerry 312

Donnelly, Maryjo xi

D'Onofrio, Richard P. 133, 212-217, 309

Dougherty, Nathan W. 33

Downey, Jane 189

Drexel Institute of Technology 307

Drucker, Dan 214

Duke University 98

Dunham, Louis J., Jr. 82, 215

Dyrud, Marilyn A. x, 29, 187

École Polytechnique 4

Educational Methods Division 71

Eisenberg, Elliot 150, 185, 330

Ellis, H. B. 156

Ellis, Robert W. 21

Emshousen, Fred W. 41, 140, 307

Engineering Deans Council 174, 178

Engineering Deans Institute 134

Engineering Education 12, 131, 325

Engineering Education News 182

Engineering Index 48, 143

Engineering Manpower Commission 11, 91, 155, 324

Engineering Spectrum Committee 183, 184

"The Engineering Technician" 12, 109, 318

Engineering Technology Archives 180

Engineering Technology College Council (ETCC) 10, 101, 159, 167-190, 334

Engineering Technology Committee 87, 133, 166, 304

Engineering Technology Council (ETC) x, 9, 88, 249, 313, 334

Engineering Technology Development Committee 25, 138, 137

Engineering Technology Division (ETD) x, 2, 189, 307

Engineering Technology Leadership Institute (ETLI) 2, 324

Engineering Workforce Commission 31, 189

Engineers' Council for Professional Development (ECPD) 8, 61, 240, 320

Engineers' Joint Council 115, 156, 292

English, Robert 21, 309

English Speaking Union Technical Teachers Exchange 43, 122, 320

Epting, Luther 21

Ethics Committee 158, 200

Faig, John T. 305

Fairleigh Dickinson University 38, 87

Fairmont State College 38

Faulkner, Steven W. 329

Federal Liaison Committee 182, 183

Federation of Regional Accrediting Commissions on Higher Education 281

Feist, K. W. 222

Fenninger, William N. 76

First National Television School 74. See also Central Technical Institute

Flaherty, Mary Ellen xi

Fleckenstein, Edward L. 125, 158, 313

Florida International University 88

Ford Foundation 69

Forman, James 14, 140, 314

Foster, C. L. 73, 156, 246

Foundation for Instrumentation Education and Research 114, 119

Francis, Lyman L. 86, 164

Franklin Institute xi, 59, 212, 307

Franklin University 312

Fraser, Gary T. 139, 165, 308

Frederick J. Berger Award 148, 185

Frederick J. Berger Award Committee 99

Freund, C. J. 306

General Education Committee 121

Gentry, Don 185

Georgia Institute of Technology 72, 203, 228

Georgia Southern College 38

Gershon, Joseph J. 85, 156, 246

Gilkeson, Mack 178

Gillespie, J. J. 118

Goals and Activities Committee 23, 143, 335

Goodson, Carole E. 218-221, 330

Gottsman, Earl E. 97, 151, 181, 307

Gourley, Frank A., Jr. 103, 133, 307

Gowdy, Kenneth K. 21, 172, 312

Graduate Studies Committee 181, 183

Grainey, Maurice R. 34, 119, 236

Greenburg, Joe S. 21

Greenwald, Stan 174

Grigsby, R. L. 309, 312

Grinter, Linton E. 26, 54, 155, 285

Grinter Report 24, 86, 215, 284

Gulley, Arnold 164

Gyte, Millard 238

Hager, Wayne 189

Hales, James A. 21, 147

Hallman, John 162, 170

Hamil, Dana B. 308

Hamme, Gary L. 21

Hammond, Harry P. 18, 61, 104, 200, 259

Harper, Arthur C. 65, 306

Harrison, Cecil 331

Hartley, H. W. 309

Hartly, Herb 118

Hartung, Walter M. 74, 108, 195, 309

Harvard University 82, 212, 232, 292

Hata, David 149, 150

Hays, Robert W. 127, 155, 205

Heiden, C. Howard 148, 331

Henninger, G. Ross 24, 80, 114, 246

Hildreth, Eddie 138

Hill, Warren R. 151, 311, 329

Historical Center Committee 138

Historical Research Committee 163

Hoffman, Larry D. 21

Holderman, Kenneth L. 70, 246

Holmes, William 50

Houston, Carl P. 21

Houze, R. Neal 21

Huband, Frank 189

Huffman, Durward R. 4, 44, 135, 312

Hughes, Walter L. 59, 104, 214, 307

Hull, Dan 173

Hutchinson, James 219

Hutton, Chuck 244

Illinois Institute of Technology 83, 125, 270

Index **359**

Independent and Private Colleges Committee 158

Indiana Northern University 90

Indiana State University, Evansville 3, 37, 91, 310

Indiana University 90, 239

Indiana University, Kokomo 38, 245

Indiana University-Purdue University, Fort Wayne 41

Indiana Vocational Tech College 38

Industrial Engineering Division 71

Institute for Electrical and Electronic Engineers (IEEE) 31, 74, 100, 149, 321

Institute for the Certification of Engineering Technicians (ICET) 84, 123, 304, 322

International Committee 188

International Journal of Applied Engineering 51

Iowa State College 113, 116

Iowa State University 131, 233

Jackson, Dugald C. 306

John Wiley & Sons Publishing Company 141, 185

Johnson, A. P. 106

Johnson, Frank 209

Johnson, Lawrence V. 72, 113, 194, 228, 308

Johnson, Martha J. 21

Joint Committee of ASEE and ACM 156

Jones, Charles S. 67, 199, 225

Jones, Russel C. 43, 44

Journal of Engineering Education 29, 127, 171, 304

Journal of Engineering Technology (JET) ix, 2, 12, 21, 28, 92, 139, 177, 326, 335

Journal Study Committee 46, 137

Junior College Division 11, 104

Junior Engineering Technology Society 285

Kanneman, Thomas A. 14, 136, 171, 312

Kansas College of Technology 329

Kansas State College 39

Kansas State Teachers College 62

Kansas State University 312

Kansas Technical Institute 14, 91, 147, 187, 308

Kerr, Howard H. 117, 307

Kirklin, B. C. 223, 227

Kirkpatrick, Edward T. 44, 95, 173, 308, 314

Koerper, Erhart 272

Kolbe, P. R. 306

Kozak, Michael R. 15

Lawshe, Chuck 237

Lehigh University 61

Lewellyn, Fred 138

Lewis and Clark College 256

Library Affairs Committee 25, 158

Lisack, J. P. 242

Lohmann, Melvin R. 78, 157, 227, 288

Long-Range Planning Committee 155, 177, 320

Louisiana State University 38, 39

Malone, James V. 130, 312

Manhattan Project 268, 269

Mankato State University 39

Mann Report 273, 285

Manpower Committee 98, 156, 324

Manpower Studies Committee 107, 332

Marcus, Joe 213

Marlow, Don 165

Marsh, R. W. 106

Martin, John R. 129, 225, 309

Maryland Institute 5

Massachusetts Institute of Technology 4, 63, 295

Masser, M. 220

Mazzola, Michael C. 88, 126, 196, 307

McBrayer, James D. 142, 312, 331

McCallick, Hugh E. 77, 123, 195, 246

McCallick Report 227

McClure, Hoyt L. 161, 165, 309

McCord, R. E. 308

McCormick, Ernie 237

McCurdy, Lyle B. 14, 30, 147, 309

McDonough, John J. 312

McGraw Award 2, 31, 59, 74, 128, 175, 181, 202, 213, 214, 218, 275, 320, 323, 332

McGraw Award Committee 84

McGraw, James H. 59, 214

Index **361**

McGraw, James L. 34

McGraw Report 25, 75

McGraw, Walter 214

McGraw-Hill Book Company xi, xii, 8, 24, 37, 87, 105, 215, 273

McHenry, Albert L. 185, 309, 311

McKay, G. 227

McNelly, George W. 195, 235

Mechanics Institute 309

Mehrhoff, Joseph C. 21

Meier, Paul T. 312

Membership Committee 113, 158, 181, 332

Membership Policy Committee 187

Memphis State University 37, 39

Merkel, Kenneth G. 4, 48, 138, 330

Merrill Publishing 312

Metz, Donald C. 84, 108, 156, 246

Michaels, Ken 241

Michigan State University 129, 260

Mid-Mississippi Valley Technology Program Directors 3, 37

Miller, Jeanne 105, 118

Miller, Kenneth R. 112

Miller, Linda C. 312

Milwaukee School of Engineering 69, 225, 270, 307, 310

Minority Action Committee 167, 178

Minority Committee 172, 181

Mississippi State University 39

Moore, Herb 163

Moore, Richard M. 21, 331

Morgan Institute of Connecticut 200

Morgulis, Issac A. 13, 131, 286

Morrill Acts 5

Morris, Sarah xi

Moss, Dorsey 244

Murray State University 13

Nashville State Technical Institute 39, 135, 312, 329

National Commission on Accreditation 115, 155, 283

National Conference on Higher Education 110

National Congress of Engineering Education 42

National Council of Technical Schools 69, 160, 198, 273

National Forum for Engineering Technology (NFET) 42, 321

National Institute for the Certification of Engineering Technologists (NICET) 3, 304

National Science Foundation (NSF) 12, 116, 292

National Society of Professional Engineers (NSPE) 69, 123, 161, 282, 321

National Study Committee 114

Neatherly, Ray 185

Neeley, V. E. 124

New Jersey Institute of Technology 38, 309

New York City University 249

New York State Institute of Applied Arts and Sciences 74, 106. See also Broome Technical Community College

New York University 74, 87, 196

Newman, William S. 312

Newton, William S. III 312

Nominating Committee 98, 158

Nominations Committee 151, 332, 333, 334, 335

North Carolina Department of Community Colleges 13, 28, 307

Northeastern University 331

Northern Arizona State University 38

Northern Arizona University 161

Northern Maine Technical College 330

Northrop Institute of Technology 309

Northwestern University 69

O'Hair, Michael T. x, xi, 4, 45, 142, 279, 312, 313

Ohio College of Applied Science 80, 225

Ohio Mechanics Institute 4, 81, 305, 311

Ohio State University 65, 155, 203

Oklahoma A & M University 107, 238, 307

Oklahoma State University 69, 77, 227, 312

Old Dominion University 307, 329

Oregon Institute of Technology xi, 45, 189, 256, 329. See also Oregon Tech- nical Institute

Oregon Technical Institute 126, 258, 312. See also Oregon Institute of Technology

Index **363**

Outstanding Service Award Committee 145

Pace, Clint 110

Paré, Ronald C. 21, 141, 309

Parks College of St. Louis University 309, 310

Parnell, Dale 183

Peer Review Process Committee 147

Pennsylvania State College 104

Pennsylvania State University xi, 64, 292, 288

Pennsylvania State University, Berks 15

Pennsylvania State University, Capital 312

Pennsylvania State University, Harrisburg 312

Pennsylvania State University, Hazleton 330

Peters, Stanton 38

Policy Committee 154

Polytechnic Institute of Brooklyn 61

Potter, A. A. 239

Pounds, J. Dale 21

Pratt Institute 65, 109, 306

Prevost, Ray G. 309

Prism 188, 304

Pritchett, Samuel L. 3, 21, 140, 329

Program Committee 126, 134, 334

Program of Work Committee 19, 136, 322

Publications Committee 25, 134, 324

Purdue University 5, 122, 227, 309

Purdue University, Anderson xi, 268

Purdue University, Calumet 38, 238

Purdue University, Fort Wayne 64, 239

Purdue University, Indianapolis 64, 240

Purdue University, Kokomo xi, 235, 322, 331

Purdue University, Michigan City 64, 240

Purvine, Winston D. 77, 125, 195, 246

QEEP 25, 142, 327

Queensboro College of New York 126

Rainey, Gil 243, 247

Rainey, Paul E. 186, 309

Rath, Gerald A. 21, 136, 307

REETS 3, 25, 133, 165

Reeve, W. Frank 330

Reid, Robert L. 42, 136

Relations with Educational Organizations Committee 115, 333, 334

Relations with Government Agencies Committee 113, 121

Relations with Government Organizations Committee 158

Relations with High Schools Committee 156, 333

Relations with Industry Committee 105, 332

Relations with Industry Division 20, 98, 279

Relations with Professional Societies Committee 107, 158, 332

Rend Lake College 38, 39

Rensselaer Polytechnic Institute 4, 132

Research and Graduate Study Directory 187

Research Committee 25, 175

Resources Committee 174

Reyes-Guerra, David R. 42, 281

Rezak, William D. xi, 21, 312, 329

Richland TEC 309

Rietzke, Eugene H. 72, 199

Rigas, Anthony L. 21

Roach, Hal 184

Rochester Athenaeum and Mechanics Institute 75. See also Rochester Institute of Technology

Rochester Institute of Technology 14, 76, 137, 172, 307. See also Rochester Athenaeum and Mechanics Institute

Rockmaker, Gordon xi

Rodes, Harold T. 60, 104, 311

Roney, Maurice W. 117, 312

Rouillion, Louis H. 309

Roulette, William 223

Roundtree, H. C. 125, 240, 307

Rudnick, Diane 14, 30, 315

Rutgers University 75

Ruth, Leo 41

Ryerson Institute of Technology 112, 152, 307

Ryerson Polytechnic Institute 13, 131, 309

Sackett, Robert L. 306

Index **365**

Samaras, Patricia 47

Sams, Denver 242

Schallert, William F. 141, 143, 309

Scholl, Lloyd W. 308

Schwake, Sara xi

Scott, Charles F. 305, 306

Scott, Ronald E. 4, 46, 331

Section Activities Committee 126, 127, 128

Self-Study Committee 127

Sharp, Dan 15

Shaw, Walter xi, 125, 308

Sims, A. Ray 73, 89, 219, 309

Sisson, Ray L. 41, 96, 141, 308

Site Selection Committee 40

Sitterlee, Lawrence J. 308

Slater, Lloyd 114

Smith, Ann Montgomery 142, 180, 312

Smith, Ellison W. 135

Smith, Eugene W. 83, 155, 308

Smith, Leo F. 307

Society for the Promotion of Engineering Education (SPEE) 9, 61, 214, 305

Society Membership Committee 138

Society of Manufacturing Engineers (SME) 50, 87, 218, 324

Society Publications Committee 148

Southern College of Technology xi, 3, 203, 210, 308. See also Southern Technical Institute

Southern Technical Institute 3, 127, 307. See also Southern College of Technology

Southwest Minnesota State College 86

Spahr, Robert H. 35, 62, 306

Sparks Award 327

Special Interest Groups Committee 142

Special Projects Committee 157

Spectrum Report 2

Speicher, Ann Lee 184

Spille, Jack 134, 163, 313

Spivey, W. T. 306

Spring Garden College 4

Sputnick 215, 258, 286

St. Louis Community College at Florissant Valley 92, 312, 331

Stanford University 295

Stanton, Frank xii, 50

State Technical Institute at Memphis 38

Stewart, Hank 140

Stocker, Don 180, 182

Stone, Ernest 130, 137

Student Selection and Guidance Committee 106, 118

Study the Fund Committee 112

Sub-committee on Divisional Reorganization 104

Sub-committee on Technical Institutes 119

Suffolk University 14

SUNY, Alfred 308

SUNY, Buffalo 106

Symposium on Dual Programs in Engineering and Engineering Technology 161

Tate, Boyce D. 38, 329

Tau Alpha Pi ix, xi, 3, 31, 148, 207, 323

Taylor, Vern 171

Teacher Training and Recruitment Committee 119

Teacher Training Committee 107

Technical College Council (TCC) 10, 63, 134, 156, 213, 333

Technical Education News (TEN) 8, 105, 111, 273, 305

Technical Institute Administrative Council (TIAC) 10, 76, 124, 313

Technical Institute Council (TIC) 10, 85, 124, 198

Technical Institute Division (TID) 8, 22, 59, 198, 273, 321

Technical Institute Studies Committee 105

Technology Accreditation Commission (TAC) 40, 91, 136, 166, 185, 304

"Technology Education Comments" 25, 137, 162

Temple University 127, 313

Texas A & M University 5, 50, 137, 169

Texas College of Mines and Metallurgy 76

Texas State Tech 157

Texas Tech University 38, 131

Texas Western College 78

Thomas, Charles 54

Thomas, Walter E. 89, 236, 164, 244, 307

Thompson, Arthur 174, 199

Thompson, Robert H. 127, 128

Tilmans, Anthony L. 3, 21, 92, 139, 307, 329

Timblin, George A. 99, 142, 307, 310

Todd, James P. 13, 162, 309

Training and Recruitment Committee 116

Travis, Harris T. 13, 94, 135, 283, 309

Trolsen, Thor 122

Troxler, G. William 1, 183, 185

Tufts University 292

Tyrrell, Cecil C. 75, 106, 124, 215

U.S. Department of Labor 108, 110

U.S. Office of Education 110, 120, 228

U.S. Technical Delegation to the Soviet Union 197

Ungrodt, Richard J. xii, 20, 79, 127, 246, 265, 268, 307

United States Naval Academy 130

University of Akron 38

University of Alabama 39

University of Arkansas, Little Rock 39

University of British Columbia 134, 161

University of California 308

University of California, Berkeley 115

University of California, Los Angeles 60, 129

University of Central Florida 331

University of Central Florida, Brevard 309, 311

University of Chicago 5, 84

University of Cincinnati 50, 144, 168, 181, 313

University of Colorado 96

University of Dayton 15, 84, 107, 124, 238, 311

University of Florida 106, 108, 117, 155

University of Houston xi, 48, 73, 199, 218, 307, 329

University of Illinois 65, 107, 116, 312

University of Iowa 78

University of Kentucky 62, 123

University of Maine 15, 70, 87

University of Maine, Orono 312

University of Maryland 39, 62

University of Massachusetts 136, 165

University of Minnesota 79

University of Missouri 86

University of Nebraska 50, 148

University of Nebraska, Omaha 48, 330

University of North Carolina 100

University of North Texas 15

University of Northern Colorado 95

University of Pennsylvania 125

University of Petroleum and Minerals 277

University of Pittsburgh 79, 92

University of Portland 147

University of Southern California 171

University of Southern Colorado 96, 308

University of Southern Illinois 37

University of Southern Mississippi 331

University of Tennessee, Knoxville 133

University of Tennessee, Martin 39

University of Texas 60, 77

University of Toledo 93

University of Virginia 84

University of Washington 60, 104

University of Wisconsin 67

Upthegrove, Bill 214

Vanderbilt University 81

Vanleer, Blake 204, 207

VanZeeland, Fred 272

VeerKamp, N. B. 312

Vermont Technical College 13, 91

Virginia Polytechnic Institute & University 39

Vocational Education Act 241, 249

Wald, Michael 51

Walk, Steve 15

Walker, Eric A. 70, 196, 292

War Training Act 239

War Training Bureau 71

Washington State University 228

Wear, Robert J. 59, 85, 128, 164, 246, 307

Weatherly, James G. 13, 30

Weber State College 311

Weidhaas, Ernest R. 87, 161, 307, 310

Welsh, William A. 144, 145, 312,

Went, L. P. 120

Wentworth Institute of Technology 28, 49, 93, 109, 143, 171, 192, 246, 258, 307, 329

Werwath, Karl O. 68, 118, 198, 215, 246, 307

West Virginia Institute of Technology xi, 7, 103

Western Kentucky University 37, 39

Westland College 91

Wichita State University 307, 312

Wickenden and Spahr Report 2, 24, 257, 273

Wickenden, William E. 35, 61, 273, 305

Wiley Award 144, 147, 180, 181, 187

Wilhelm, William 21

Willenbrock, Karl 184

Williams, Ron 134, 137, 169, 178

Williamson, Merritt A. 81, 246

Williston, Arthur L. 63, 109, 124

Williston Award 158, 320, 333

Wisz, Ted 173

Wolf, Barbara A. 47, 51, 52, 330

Wolf, Lawrence J. xi, 4, 43, 93, 169, 330

World Conference on Applied Engineering and Engineering Technology 143

World Conference on Engineering and Engineering Technology 235

World Congress on Education in Engineering and Engineering Technology 141

World Congress on Engineering Technology Education 327, 328

Worthley, Warren W. 41, 42

Wyomissing Polytechnic Institute 65

Yale University 80

Zanetti, Peter 180

MILESTONES OF ENGINEERING TECHNOLOGY EDUCATION

1931 WICKENDEN AND SPAHR STUDY OF TECHNICAL INSTITUTES RECOGNIZES THE PLACE OF ENGINEERING TECHNOLOGY EDUCATION IN THE TECHNICAL SPECTRUM **1941** THE SOCIETY FOR THE PROMOTION OF ENGINEERING EDUCATION (SPEE) ESTABLISHES TECHNICAL INSTITUTE DIVISION (TID) ❑ MCGRAW-HILL BOOK COMPANY PUBLISHES FIRST *TECHNICAL EDUCATION NEWS* **1945** ENGINEERING COUNCIL FOR PROFESSIONAL DEVELOPMENT (ECPD) ESTABLISHES FIRST ACCREDITATION PROCEDURES FOR TWO-YEAR PROGRAMS **1946** FIRST ASSOCIATE ENGINEERING TECHNOLOGY PROGRAM ACCREDITED ❑ SPEE BECOMES THE AMERICAN SOCIETY FOR ENGINEERING EDUCATION (ASEE) ❑ TID REORGANIZED — COMMITTEE OF 21 FORMED **1950** FIRST JAMES H. MCGRAW, SR., AWARD PRESENTED **1953** TAU ALPHA PI HONORARY SOCIETY ESTABLISHED **1954** "THE ENGINEERING TECHNICIAN" PAMPHLET PUBLISHED **1955** L.E. GRINTER REPORT: " REPORT OF THE COMMITTEE ON EVALUATION OF ENGINEERING EDUCATION" PUBLISHED **1959** *JOURNAL OF ENGINEERING EDUCATION* NOVEMBER ISSUE FEATURES ENGINEERING TECHNOLOGY EDUCATION ❑ HENNINGER STUDY, *THE TECHNICAL INSTITUTE IN AMERICA*, PUBLISHED **1962** ASEE REORGANIZED AND THE TECHNICAL INSTITUTE COUNCIL (TIC) FORMED ❑ MCGRAW REPORT: "CHARACTERISTICS OF EXCELLENCE IN ENGINEERING TECHNOLOGY EDUCATION: THE EVALUATION OF TECHNICAL INSTITUTE EDUCATION" PRESENTED **1964** *JOURNAL OF ENGINEERING EDUCATION* JULY-AUGUST ISSUE FEATURES ENGINEERING TECHNOLOGY EDUCATION **1965** FIRST B.S. ENGINEERING TECHNOLOGY PROGRAM ACCREDITED ❑ TIC BECOMES TECHNICAL INSTITUTE ADMINISTRATIVE COUNCIL (TIAC) ❑ TID ESTABLISHES FUNCTION OF SECTION REPRESENTATIVES **1966** *JOURNAL OF ENGINEERING EDUCATION* NOVEMBER ISSUE FEATURES ENGINEERING TECHNOLOGY EDUCATION **1971** TIAC BECOMES ENGINEERING TECHNOLOGY COLLEGE COUNCIL (ETCC)❑